P9-DUC-735

Bayes or Bust?

Bayes or Bust?
A Critical Examination of Bayesian Confirmation Theory

John Earman

A Bradford Book
The MIT Press
Cambridge, Massachusetts
London, England

This book was set in Times Roman by Asco Trade Typesetting Ltd., Hong Kong, and was printed and bound in the United States of America.

Library of Congress Cataloging-in-Publication Data

Earman, John.
   Bayes or bust? : a critical examination of Bayesian confirmation theory / John Earman.
      p.   cm.
   "A Bradford book."
   Includes bibliographical references and index.
   ISBN 0-262-05046-3
   1. Science—Philosophy. 2. Science—Methodology. 3. Bayesian statistical decision theory. I. Title.
Q175.E23   1992
502.8—dc20   91-35623
                                                                              CIP

For Peter Hempel, teacher extraordinaire,
And Grover Maxwell, a gentleman who left his mark on an
ungentlemanly world

Contents

Preface

Philosophers of science can be justifiably proud of the progress achieved in their discipline since the early days of logical positivism. A glaring exception concerns the analysis of the testing of scientific hypotheses and theories, an exception that threatens to block further progress toward a central goal not only of the logical positivists but also of their predecessors and heirs. That goal is the understanding of "the scientific method." Whatever else this method involves, its principal concern is with the issue of how the results of observation and experiment serve to support or undermine scientific conjectures. Not only does contemporary philosophy of science fail to provide a persuasive analysis of this issue, there are even rumblings to the effect that a search for an "inductive logic" or "theory of confirmation" is a fruitless quest for a nonexistent philosopher's stone. The principal difficulty here, it should be emphasized, does not derive from Kuhnian incommensurability and its fellow travelers, for the issue has resisted resolution even for cases that do not stray near the frontiers of scientific revolutions.

This work explores one dimension of this impasse by providing a critical evaluation of the approach I take to provide the best good hope for a comprehensive and unified treatment of induction, confirmation, and scientific inference: Bayesianism. It is intended for students of the philosophy of scientific methodology, where 'student' is used in the broad sense to include advanced undergraduates, graduate students, and professional philosophers. It is also intended to annoy both the pro- and contra-Bayesians. To fling down the gauntlet, the critics of Bayesianism have generally failed to get the proper measure of the doctrine, while the Bayesians themselves have failed to appreciate the pitfalls and limitations of their approach. And to add insult to insult, neither side has appreciated the source of the doctrine—the Reverend Bayes's essay. Nor is there much appreciation of what the new discipline of formal learning theory has to tell us about the latent assumptions responsible for the apparent reliability of Bayesian methods. If the annoyance serves as a spur to further progress, I will count this book a success.

I want to emphasize as strongly as I can that this work is *not* intended as a comprehensive review of the pros and cons of Bayesianism. Issues and points of views on issues have been selected with an eye to giving the reader a sense of the strengths and weaknesses of the Bayesian approach to confirmation, and my selections should be judged on that basis.

The reader who penetrates very far into this work may begin to experience a topsy-turvy feeling. Those who complain will in effect be crying, Stop the world, I want to get off! For the world of confirmation is a topsy-turvy world.

My intellectual debts are too numerous to recount here. But I would be remiss if I did not acknowledge the many helpful suggestions I received on earlier drafts of this book. Thank you Jeremy Butterfield, Charles Chihara, Alan Franklin, Donald Gillies, Clark Glymour, David Hillman, Colin Howson, Richard Jeffrey, Cory Juhl, Kevin Kelly, Philip Kitcher, Tim Maudlin, John Norton, Teddy Seidenfeld, Elliott Sober, Paul Teller, Jan von Plato, Bas van Fraassen, and Sandy Zabell. With all of this help, the book should be better than it is.

The material in chapter 1 appeared as "Bayes' Bayesianism," *Studies in the History and Philosophy of Science* 21 (1990): 351–370. A version of chapter 5 appeared as "Old Evidence, New Theories: Two Unsolved Problems in Bayesian Confirmation Theory," *Pacific Philosophical Quarterly* 70 (1989): 323–340. I am grateful to the editors of these journals for their permission to reprint this material. The excerpts from Bayes's essay found in the appendix to chapter 1 are reproduced with the kind permission of the editors of *Biometrika*.

Notation

Logic

$(\forall x)$	universal quantifier
$(\exists x)$	existential quantifier
&	conjunction, A & B (A and B)
$\&_{i \leqslant n} A_i$	A_1 & A_2 & ... & A_n
$\vee$	disjunction, $A \vee B$ (A or B or both)
$\bigvee_{i \leqslant n} A_i$	$A_1 \vee A_2 \vee ... \vee A_n$
$\rightarrow$	material implication, $A \rightarrow B$ (if A then B)
$\leftrightarrow$	biconditional, $A \leftrightarrow B$ (A if and only if B)
$\neg$	negation, $\neg A$ (not A)
$\equiv$	definition
$\models A$	A is valid (i.e., A is true in every possible world or model)
$A \models B$	A semantically implies B (i.e., B is true in every possible world or model in which A is true)

Mathematics

$x \in X$	x is an element of X
$X \subseteq Y$	X is a subset of Y
$X \cap Y$	the intersection of X and Y
$X \cup Y$	the union of X and Y
$\sum_i$	summation over the index i
$\int$	integral
$\mathbb{N}$	natural numbers
$\mathbb{R}$	real line
$f: X \rightarrow Y$	f maps $x \in X$ to $f(x) \in Y$
sup	supremum
$\lim_{n \to \infty}$	the limit as n approaches infinity

Probability and Statistics

$\Pr(A)$ the (unconditional) probability of A

$\Pr(A/B)$ the conditional probability of A on B

$\mathrm{E}(X)$ the expectation value of the random variable X

$\binom{n}{m}$ $n!/[m!(n-m)!]$

Bayes or Bust?

Introduction

The Reverend Thomas Bayes died in 1761, leaving a not inconsiderable estate. Among the bequests was one for £100 to Richard Price, "now I suppose preacher at Newington Green."[1] This tentative identification of Price's whereabouts indicates a lack of close contact between the two men when Bayes's will was executed in 1760. But whatever the lack of personal contact, there was a closeness of interests and outlooks: both men were nonconformist preachers, both were fellows of the Royal Society, and both were mathematicians of some accomplishment. Price discovered among Bayes's papers two essays, which he communicated to John Canton, secretary of the Royal Society. One dealt with the properties of asymptotic series, the other with what Price called "analogical or inductive reasoning." The latter was published in the 1763 volume of the *Philosophical Transactions* under the title "An Essay Towards Solving a Problem in the Doctrine of Chances."

During the remainder of the eighteenth century and for most of the nineteenth century, Bayes's essay seems to have had little influence on either the practice of statistics or the study of its foundations, and in the early twentieth century the giants of probability and statistics, such as Pearson, Fisher, and Jeffreys, read Bayes through lenses heavily tinted by their own views of probability.[2] Bayes's essay became better known after it was reprinted in facsimile in 1940 and in a new edition in *Biometrika* in 1958, but if the state of the secondary literature is any indication, it wasn't until quite recently that statisticians and philosophers have made serious attempts to understand what Bayes had written.[3] Yet even if Bayes's actual ideas have not been much used, his name certainly has. "Bayesianism" is the label stitched to the Jolly Roger of a leading school of statistics and of what is arguably the predominant view among philosophers of science concerning the confirmation of scientific hypotheses and scientific inference in general. The extent to which Bayes himself should be counted as a Bayesian in modern terms is a question tackled in chapter 1. The chapters that follow are focused mainly on the confirmation of nonstatistical hypotheses and theories. I will have little to say about technical issues in Bayesian statistics, although such issues do intrude from time to time, as with the question of countable additivity (see chapter 2).

I confess that I am a Bayesian—at least I am on Mondays, Wednesdays, and Fridays. And when I am a Bayesian I am an imperialistic apostle in insisting that every sin and virtue in confirmation theory should be explained in Bayesian terms. My Monday, Wednesday, and Friday enthusiasm has even led me to the extreme of rejecting various views on confirmation held by my revered colleagues Clark Glymour, Adolf Grünbaum, and Wesley Salmon. On Tuesdays, Thursdays, and Saturdays, however, I have my doubts not only about the imperialistic ambitions of Bayesianism but also about its viability as a basis for analyzing scientific inference. (On Sundays I try not to think about the matter.) I hasten to add that my own schizophrenia on this topic, deplorable as it may be, is symptomatic of a deep schism in the philosophical community. The Bayesians and their camp followers show an impatience tinged with contempt for those who dare to doubt the orthodoxy. But doubters there are, and even a few claim flat out that the orthodoxy is unworkable.

This is a topic I find worth pursuing not only for the sake of my own mental health but also because of its wide ramifications. Bayesianism is the only view presently in the offing that holds out the hope for a comprehensive and unified treatment of inductive reasoning. If the hope is a vain one, the better we should know as soon as possible so that we can begin work on another approach. If, on the contrary, the hope can be fulfilled, then we should all commit ourselves to labor in the vineyard planted by the Reverend Thomas. And beyond such concerns of global strategy, there is much to be learned about the confirmation of scientific hypotheses and the problem of induction by attending to the clash between the Bayesians and their critics. Whatever one's ultimate decision about Bayesian confirmation theory, it possesses the one unmistakable characteristic of a worthy philosophical doctrine: the harder it is pressed, the more interesting results it yields. I intend the chapters that follow, even the critical ones, to showcase this characteristic.

Chapter 1 begins at the beginning and traces the curious logic of Bayes's essay. This is an exercise that has an interest beyond the merely historical, for the attempt to understand Bayes in his own terms reveals a microcosm of problems that still reverberate through modern-day discussions of the foundations of probability and inductive inference. This chapter does not assume that the reader is familiar with modern Bayesianism but does presuppose an acquaintance with the basics of probability theory. Those readers who have not already made this acquaintance may wish to consult chapter 2 first and then return to chapter 1.

Chapter 2 gives a brief review of the technical apparatus of modern Bayesianism, including the probability calculus, Bayes's theorem, and rules for changing degree of belief via conditionalization. The probability axioms adopted there are the standard ones, plus a special form of countable additivity used repeatedly in subsequent chapters. Some technical issues related to the general form of countable additivity are discussed in an appendix. The popular Dutch-book justification for the probability axioms, with probability interpreted as degree of belief, is discussed with an admittedly critical bias. But if this justification is found wanting, others are ready to take its place. Three of these alternative justifications are reviewed.

Chapter 3 is a sermonette addressed to the uninitiated and the unconverted. The discussion is designed to display the analytical prowess of Bayesianism by showing how it can be used to dissect the strengths and weaknesses of other approaches to confirmation, including hypotheticodeductivism, Hempel's instance confirmation, and Glymour's bootstrap testing. In addition, Bayesianism is shown to provide a satisfactory resolution of Hempel's infamous ravens paradox and to help to make sense of such truisms of confirmation theory as that variety of evidence can count more than sheer quantity of evidence. Further, Bayesianism is shown to provide an illuminating means of testing various claims about the indispensability of theories in scientific inference. And Bayesianism is also shown to provide the form of a solution to Quine's and Duhem's problem, though how to instantiate the form depends on a resolution of the objectivity problem taken up in chapter 6.

Chapter 4 is supposed to quiet the doubts raised by a number of critics. These critics variously charge that the Bayesian apparatus never gets into gear for scientific laws because they receive flatly zero priors (Popper); that the gears turn but to no avail, since the probabilification that hypotheses receive is never genuine inductive support (Popper and David Miller); that the gears turn too easily and too fast, which yields ersatz confirmation (Grünbaum); and that the turning of the gears is accompanied by a nasty grinding sound because the teeth get snagged on the problem of adhocness that vitiated the hypotheticodeductive method (Richard Miller). The Bayesian approach is also shown to meet the challenge of Nelson Goodman's new problem of induction; indeed, it is argued that without the help of the Bayesian apparatus, it is hard, if not impossible, even to state Goodman's problem in a precise and persuasive form. The verdict on the Bayesian analysis of the importance of novelty of prediction is somewhat more equivocal.

Having discussed the origins of Bayesianism, exposited the tenets of its modern form, extolled its virtues, and defended it against its leading critics, I then turn the proceedings over to my alter ego, who takes a more jaundiced view of Bayesianizing. Chapter 5 takes up the problem of "old evidence"—the seeming inability of the personalist form of Bayesianism to account for the confirmatory power of evidence already known to be true. The source of the interesting and recalcitrant version of the problem is traced to the failure of scientists to be logically omniscient in the sense that they are unable to parse in advance all of the possibilities in the space of hypotheses. Garber, Jeffrey, Niiniluoto, and others have attempted to solve the problem by exploiting a second aspect of the failure of logical omniscience, the failure of real scientists to immediately recognize logical and mathematical implications. They allow for a more humanized form of Bayesianism in which the agents learn logical-mathematical truths and change their degrees of belief via conditionalization on this new nonempirical knowledge. I show, however, that while such an allowance is highly desirable, it does not adequately resolve the original version of the old evidence problem. Other attacks on the problem are discussed, and all are found wanting.

Chapter 6 examines the two main attempts to provide a Bayesian explanation of the rationality and objectivity of scientific inference: first, through the objectification of prior probabilities and, second, by "washing out" priors with accumulating evidence. Neither attempt stands up to scrutiny. The prospects for other Bayesian explanations of objectivity are assayed and found wanting. These results can be read in two radically different ways: the Bayesian account of confirmation is inadequate or else the vaunted objectivity of science is open to serious question.

Chapter 7 is a plea for what is widely regarded as an outmoded and discredited view of scientific inference, eliminative induction. I argue that an interesting form of Bayesian inductivism can hardly succeed without being eliminativist. However, the form of eliminativism recommended here is not the quaint Sherlock Holmes variety that would turn induction into deduction but a more sophisticated version that joins hands with elements of Bayesianism. The self-conscious effort at eliminative induction undertaken in recent years in relativistic gravitational theory is analyzed in some detail. This case study emphasizes the need for the active exploration of the space of possible hypotheses, a need that derives from the failure of logical omniscience discussed in chapter 5.

Chapter 8 brings together Thomas Bayes and Thomas Kuhn to see what each can learn from the other. Kuhn's points that in scientific revolutions there is no neutral algorithm for theory choice and that persuasion rather than proof is the order of the day find a sympathetic resonance in the humanized form of Bayesianism that recognizes nonconditionalization shifts in degrees of belief that occur when new possibilities are introduced. Such shifts, it is argued, are not governed by any neat formal rules. But more generally, Kuhn's conceptualization of theory choice and his insistence that it is the community of experts rather than the individual members that makes the choice are shown to be incommensurable with Bayesianism. Here I side largely with Tom Bayes rather than with Tom Kuhn.

Chapter 9 begins with an examination of Putnam's diagonalization argument against Carnap's systems of inductive logic. The argument is important because, if correct, it tells against Bayesianism in general. It is also important because it led Putnam to an alternative conceptualization of inductive inquiry that in the last few years has begun to bear fruit under the label of formal-learning theory. This development raises in a new form some of the old questions about the logic of discovery and the relationship between discovery and justification. Formal-learning theory also helps to unmask some of the disconcerting presuppositions built into the convergence-to-certainty results needed to link Bayesian personalism to truth and reliability. This unmasking makes the Bayesian look embarrassingly like a dogmatist.

The upshot of my examination of Bayesian confirmation theory is neither a simple thumbs up nor a simple thumbs down. Chapter 10 attempts to present a more equivocal assessment with the help of a Galilean dialogue.

1 Bayes's Bayesianism

1 Bayes's Problem

The title Price gave to Bayes's essay, "An Essay Towards Solving a Problem in the Doctrine of Chances," gives no real clue as to its groundbreaking aims. Up to the time of Bayes's essay, most of the work in the doctrine of chances had concerned what can be called direct inferences: given the basic probabilities for some chance setup, calculate the probability that some specified outcome will occur in some specified number of trials, or calculate the number of repetitions needed to achieve some desired level of probability for some outcome; e.g., given that a pair of dice is fair, how many throws are needed to assure that the chance of throwing double sixes is at least 50/50?[1] Bayes's essay was concerned with the problem of inverse inference: given the observed outcomes of running the chance experiment, infer the probability that this chance setup will produce in a given trial an event of the specified type. More exactly, the problem that Bayes set himself was this: "*Given* the number of times in which an unknown event has happened and failed: *Required* the chance that the probability of its happening in a single trial lies somewhere between any two degrees of probability that can be named" (p. 298).[2] Significantly, Bayes thought that just as the answer to a problem of direct probabilities was to be given by a calculation leading to a definite numerical value, so the solution to his problem had to be supplied in the same terms.

Bayes's essay may, for sake of analysis, be divided into four parts. Part 1 provides a definition of probability, which is then used to demonstrate various principles of probability now taken as basic axioms or theorems of the calculus of probability (Bayes's propositions 1 to 7). These principles are put to work in part 2 to derive formulas relating to a chance setup utilizing a perfectly level billiard table (Bayes's propositions 8 and 9 and their corollaries).[3] Part 3 consists of a scholium in which Bayes attempts to justify taking the formulas from part 2 as providing the form of a solution to his problem. Part 4 (Bayes's proposition 10, the rules, and an appendix)[4] completes the solution by providing concrete applications and numerical estimates of the integrals occurring in parts 2 and 3.[5] My main focus will be on parts 2 and 3, which contain the key arguments for Bayes's proposed solution to his problem. But before I turn to these arguments, it is necessary to do some initial spade work on part 1. For the reader's convenience, Bayes's main definitions and propositions are collected in an appendix to this chapter.

2 Bayes's Two Concepts of Probability

A central difficulty that surfaces in Bayes's attack on his problem can be traced to a tension that arises from trying to combine two concepts of probability. At the risk of introducing misleading connotations, I will refer to these two concepts as *degree of belief* and *objective probability*. The first concept is set out in definition 5 of Bayes's essay: "The *probability of any event* is the ratio between the value at which an expectation depending on the happening of the event ought to be computed, and the value of the thing expected upon its happening" (p. 298). Modern Bayesian personalists may wish to see this definition as the touchstone of their conception of probability, and with the omission of only a couple of words the definition can be given a personalist gloss: the probability of an event for a given agent is the maximum amount that the agent is willing to spend for a contract that pays (say) £1 if the event occurs and nothing otherwise. But Bayes's qualifying phrase "ought to be" suggests that what he had in mind was not subjective or personal degree of belief but something more akin to rational or justified degree of belief. This suggestion is supported by the fact that the solution Bayes offers to his problem supplies determinate numerical values not subject to the whims of different agents.

Bayes's position also differs from that of thoroughgoing Bayesian personalists who do not recognize any notion of objective chance apart from a limiting notion of objectified degree of belief. In his preface Price praises De Moivre's work for helping to show that "there are in the constitution of things fixt laws according to which events happen, and that, therefore, the frame of the world must be the effect of the wisdom and power of an intelligent cause; and thus to confirm the argument taken from final causes for the existence of the Deity" (p. 297). But Bayes deserves even higher praise, because "the converse problem solved in this essay is more directly applicable to this purpose; for it shews us, with distinctness and precision, in every case of any particular order or recurrency of events, what reason there is to think that such recurrency or order is derived from stable causes or regulations in nature, and not from any of the irregularities of chance" (p. 297). Some modern readers will interpret Price's "order or recurrency of events" as the tendency of the chance setup to yield stable frequencies, and his "stable causes or regulations" as physical propensities that are inherent in the chance setup and responsible for the observed frequencies in repeated trials. Whether or not Bayes would have endorsed such a reading is impossible to determine with any certainty from the text of his essay, but the application given in rule 1 suggests the interpretation of a single-case propensity (see section 7 below). In any case, the very statement of his problem hardly makes sense unless read as calling for a rational estimate of the fixed but unknown value of a nonepistemic, objective chance, whether interpreted in terms of propensities, in terms of limiting frequencies, or in some entirely different manner.

The difficulty here is not that Bayes failed to give a definition of 'objective probability'. In modern parlance, 'objective probability' is a theoretical term like 'spin' and 'charm'. Such terms do not stand in need of any definition over and above the implicit definitions they receive from the roles they play in the theory—the theory of elementary particles in the case of 'spin' and 'charm', the theory of chance setup in the case of 'objective probability'. Rather, the difficulty for Bayes comes in talking about the degree of belief in the value of the objective probability. Recall that for Bayes probability qua degree of belief is assigned to events, and whatever the metaphysical status of events, they are things that can be ascertained to happen or fail to happen, since otherwise the expectation or contract whose economic value defines the degree of belief cannot be paid off.[6]

The most obvious response to this difficulty is to find a proxy event that will do duty for the state of affairs of the objective probability's lying between specified limits. That this is the course actually followed by Bayes is made abundantly clear by part 2 of his essay, as we will see below in section 6.

3   Bayes's Attempt to Demonstrate the Principles of Probability

Price informs us that "Mr Bayes has thought fit to begin his work with a brief demonstration of the general laws of chance. His reason for doing this, as he says in his introduction, was not merely that his reader might not have the trouble of searching elsewhere for the principles on which he has argued, but because he did not know whither to refer him for a clear demonstration of them" (p. 298). Given Bayes's definition of probability, the modern reader might expect to find that Bayes's demonstration of the laws of chance follows the strategy of showing that a violation of any of the laws would lead to Dutch book (see chapter 2). In fact, what Bayes does is both more interesting and more problematic.

An example of Bayes's style of reasoning is given by his proof of proposition 1, the principle of finite additivity for "inconsistent events." Let A, B, C be pairwise incompatible events, and suppose that your probabilities for these events are respectively a, b, and c. If we let $[\pounds P; Z]$ stand for the expectation or contract that awards payoff P, contingent on the occurrence of the event Z, these probability assignments mean that you are willing to spend $\pounds a$ for $[\pounds 1; A]$, $\pounds b$ for $[\pounds 1; B]$, and $\pounds c$ for $[\pounds 1; C]$. Under these conditions, Bayes asserts you should be willing to spend $\pounds(a + b + c)$ for $[\pounds 1; A \text{ or } B \text{ or } C]$. So by the definition of probability, your probability for the disjunctive event A or B or C is the sum of your probabilities for the individual events. Bayes's argument here seems to be that $\text{val}[\pounds 1; A \text{ or } B \text{ or } C] = \text{val}[\pounds 1; A] + \text{val}[\pounds 1; B] + \text{val}[\pounds 1; C]$ because having the contract $[\pounds 1; A \text{ or } B \text{ or } C]$ is equivalent to having the three contracts $[\pounds 1; A]$, $[\pounds 1; B]$, and $[\pounds 1; C]$ and because the value of having the three is the sum of their values. This value additivity principle has been criticized by Schick (1986) as part of his attack on the Dutch-book justification of the probability axioms. I do not now wish to enter this dispute except, first, to reiterate Schick's point that value additivity is an implicit assumption of the Dutch-book construction and, second, to note that given value additivity, Bayes's argument already justifies the finite additivity axiom of probabilities without any need to bring in Dutch bookies.[7] But note also that Bayes's solution to his problem involves a principle much stronger than finite additivity (see sections 4 and 8 below and also chapter 4).

The corollary to proposition 1 is the negation rule, $\Pr(A) + \Pr(\neg A) = 1$, which follows immediately from proposition 1 on the assumption that $\text{val}[\pounds N; A \text{ or } \neg A] = N$ (here val is measured in pound units). This rule is used in turn to establish proposition 2, which asserts that the probability of an event is to the probability of its failure as the loss, if it fails, is to the gain, if it happens. For $\Pr(A) = P/N$ means that for me $\text{val}[\pounds N; A] = P$. This implies that my loss is $\pounds P$ if A fails while my gain is $\pounds(N - P)$ if A obtains. Hence by the corollary to proposition 1, $(\Pr(A)/\Pr(\neg A)) = (P/(N - P))$.

Propositions 3 and 5, which concern the "subsequent events" A and B, have been rendered by modern commentators respectively as

$$Pr(B/A) = \frac{Pr(B \& A)}{Pr(A)}$$
(1.1a)

and

$$Pr(A/B) = \frac{Pr(A \& B)}{Pr(B)}.$$
(1.1b)

There is a double puzzle here. First, to modern eyes, (1.1a) and (1.1b) look like definitions rather than substantive propositions needing proofs. Second, having established (1.1a) in proposition 3, it would seem that (1.1b) could be established by a parallel proof, whereas the proof Bayes actually offers for proposition 5 is entirely different from that in proposition 3. Shafer (1982) suggests that time order is the key to understanding the difference Bayes saw between (1.1a) and (1.1b), but the time order of A and B seems to me not to play any role in Bayes's proof of proposition 3.[8] Another suggestion for resolving these puzzles is as follows. As to the first puzzle, (1.1a) may be seen as a definition of conditional probability, but the definition still stands in need of justification or an operational underpinning, which is what proposition 3 is supposed to provide. As to the second puzzle, the fact that the proof of proposition 5 is so different from that of proposition 3 suggests that in proposition 5 Bayes was after more than a mere variant of proposition 3. Let $Pr_B(\cdot)$ stand for the probability that results from $Pr(\cdot)$ upon learning that B obtains. Thus, when Bayes speaks in proposition 5 of "the probability I am right" if I guess that A has happened, "it first being discovered that the 2nd event [B] has happened" (p. 301), he may have intended $Pr_B(A)$. In this notation, what proposition 5 then comes to is that $Pr_B(A) = Pr(A/B) \equiv Pr(A \& B)/Pr(B)$, or in modern jargon that learning is modeled as conditionalization. This understanding of proposition 5 is consonant with taking Bayes to be calling for the value of $Pr_{X=x}(p_1 \leqslant p \leqslant p_2)$ when he states that his problem is to find the chance that the probability p of the unknown event lies between the limits p_1 and p_2, given the number of times x in which the event has occurred in n trials. It must be admitted, however, that Bayes's intentions in propositions 3 and 5 are far from pellucid, and the text is ambiguous as to whether he meant $Pr(A/B)$ or $Pr_B(A)$ or a mixture of the two. In fact, the matter may not have any resolution, since in Bayes's day there was no well-developed concept of conditional probability and certainly no standard notation for it (see Shafer 1982).

In attempting to prove proposition 3, Bayes sets $Pr(A \& B) = P/N$ and $Pr(A) = a/N$, and he supposes that the probability that "the 2nd $[B]$ will happen upon the supposition the 1st $[A]$ does" (p. 299), i.e., $Pr(B/A)$ (or perhaps $Pr_A(B)$) is b/N. $Pr(A \& B) = P/N$ and $Pr(A) = a/N$ mean respectively that $val[£N; A \& B] = P$ and $val[£N; A] = a$. Bayes then reasons that in the contract contingent on $A \& B$ one's expectation "will become b if the 1st $[A]$ happens" (p. 299). This means, he says, that if A happens, one's gain is $b - P$, while if A fails, one's loss is P. Thus by proposition 2, he concludes that $(P/(b - P)) = (a/(N - a))$, which is then manipulated to produce the desired result. The demonstration is not convincing, since the sense of 'gain' here is not the same as that in proposition 2.

There are, however, enough materials in Bayes's essay for a more satisfying justification. As we will see shortly, proposition 5 introduces the notion of a nested contract $[[£N; C]; D]$, whose payoff is the contract $[£N; C]$ if D obtains. Two plausible principles can be adopted in evaluating such contracts. First, $val[[£N, C]; D] = val[£N; C \& D]$. Second, $val[[£N; C]; D] = Pr(D) \times val_D[£N; C]$, where $val_D[\cdot]$ stands for the value of $[\cdot]$ conditional on the supposition that D obtains. But $val_D[£1; C]$ may be taken as the definition of the conditional probability $Pr(C/D)$. And since $val[£1; C \& D] = Pr(C \& D)$, we obtain $Pr(D) \times Pr(C/D) = Pr(C \& D)$.

Since proposition 5 is crucial to parts 2 and 3 of Bayes's essay, I will examine his attempted proof in some detail. This attempt is based on proposition 4, whose curious logic has been illuminated by Shafer (1982). Let there be two subsequent events A_1, B_1 to be determined on the first day, two subsequent events A_2, B_2 to be determined on the second day, and so forth without end. The probability $Pr(B_i)$, $i = 1, 2, 3, \ldots$, for the occurrence of the second event on day i is supposed to be the same for all i, and likewise for the probability $Pr(A_i \& B_i)$ for the coincidence of the two events on day i. Let $E_j, j = 1, 2, 3, \ldots$, be the event that occurs just in case an A event happens on the first day that a B event happens starting from day j.[9] With $A_1 = A$ and $B_1 = B$, proposition 4 asserts that

$$Pr(E_1) = Pr(A \& B)/Pr(B). \tag{1.2}$$

Shafer's reconstruction of Bayes's proof of (1.2) is elegant and brief. Bayes assumes, on Shafer's account, that $Pr(E_1) = Pr(E_2)$ and also that E_2 is independent of (A_1, B_1). By construction, E_1 is the same as $(A_1 \& B_1) \vee (\neg B_1 \& E_2)$. Therefore,

$$\begin{aligned}
\Pr(E_1) &= \Pr(A_1 \ \& \ B_1) + \Pr(\neg B_1 \ \& \ E_2) \\
&= \Pr(A_1 \ \& \ B_1) + \Pr(\neg B_1) \times \Pr(E_2) \\
&= \Pr(A_1 \ \& \ B_1) + (1 - \Pr(B_1)) \times \Pr(E_2) \\
&= \Pr(A_1 \ \& \ B_1) + (1 - \Pr(B_1)) \times \Pr(E_1).
\end{aligned} \tag{1.3}$$

The first equality follows from additivity (proposition 1), the second by the multiplication principle for independent events, the third from the negation principle (corollary to Bayes's proposition 1), and the fourth from the starting assumptions. Equation (1.2) then follows by rearrangement of terms.

Bayes's own attempted proof, however, is somewhat different from Shafer's reconstruction. In particular, Bayes makes no direct appeal to the multiplication principle for independent events; indeed, this principle is not stated until afterward as proposition 6. To understand Bayes's reasoning, start from the fact that $\Pr(E_1) = x$ means that $\mathrm{val}[\pounds 1; E_1] = x$. On the first day, I have the expectation of receiving £1 if both A_1 and B_1 obtain. But if this coincidence fails, I have "an expectation of being reinstated in my former circumstances, i.e. of receiving that which in value is x depending on the failure of the second event" (p. 300). Assuming that $\mathrm{val}[\pounds 1; E_1] = \mathrm{val}[\pounds 1; E_2]$, this second expectation can be interpreted as the nested contract $[[\pounds 1; E_2]; \neg B_1]$. Since these two expectations "together are evidently the same with my original expectation," what is needed to establish proposition 4 is a valuation principle for nested contracts that sets $\mathrm{val}[[\pounds 1; E_2]; \neg B_1] = \mathrm{val}[\pounds 1; E_2] \times \Pr(\neg B_1)$, which is what Bayes seems to mean when he writes, "Wherefore since x is the thing expected and y/x the probability of obtaining it, the value of this expectation is y." From what was said above, we should have $\mathrm{val}[[\pounds 1; E_2]; \neg B_1] = \mathrm{val}[\pounds 1; E_2 \ \& \ \neg B_1] = \Pr(E_2 \ \& \ \neg B_1)$. Thus it seems that the multiplication axiom for independent events must be tacitly invoked to get $\Pr(E_2 \ \& \ \neg B_1) = \Pr(E_2) \times \Pr(\neg B_1) = \mathrm{val}[\pounds 1; E_2] \times \Pr(\neg B_1)$.

The proof of proposition 5 is completed by adjoining to proposition 4 two further equalities derived in the corollary:

$$\Pr_B(E_1) = \Pr_B(A) \tag{1.4}$$

and

$$\Pr_B(E_1) = \Pr(E_1) \tag{1.5}$$

The first of these is straightforward, since, as Bayes says, "after this discovery [that B obtains,] the probability of my obtaining N [the payoff of the contract contingent on E_1] is the probability that the 1st of the two subsequent events [A] has happened upon the supposition that the 2nd has" (p. 300). The argument for (1.5), however, is more problematic. Suppose first that (1.5) fails because $Pr_B(E_1) < Pr(E_1)$. Then before it is known whether or not the first event has happened, "it would be reasonable for me to give something to be reinstated in my former circumstances, and this over and over again as often as I should be informed that the 2nd event had happened, which is evidently absurd" (p. 300). Presumably, as Shafer (1982) offers, Bayes means that under the *reductio* supposition one should be willing to pay a premium to void the original contract [£1; E_1] in favor of a new contract [£1; E_2]; for before one learns that B happens, [£1; E_1] and [£1; E_2] have equal values, and afterward [£1; E_1] drops in value, while the value of [£1; E_2] is presumably unaffected. And this would take place over and over again with E_3, E_4, etc. each time it is learned that the B event has occurred, which leads to an indefinitely large pay out. Though somewhat reminiscent of Dutch-book arguments, the present argument lacks their bite. The unfortunate fellow caught in a Dutch book faces sure ruin, no matter what events the vicissitudes of chance choose to actualize, whereas on Bayes's construction there is only a contingent ruin in the offing based on the continued occurrence of B events. But contingent ruin is not a very powerful stick, since it threatens all too often even those who are scrupulous to obey the principles of probability. The other half of Bayes's *reductio* is even less compelling. If $Pr_B(E_1) > Pr(E_1)$, there is a premium one will refuse as an inducement to cancel [£1; E_1] in favor of [£1; E_2], and this over and over again as often as B events continue to occur. Refusing to follow a course of action that as a matter of fact would lead to unlimited gain is a matter for regret, but it does not seem to be the kind of regret on which a theory of probability should be founded.

4   Bayes's Principle-of-Insufficient-Reason Argument

In his preface Price refers to an "introduction which he [Bayes] has writ to this Essay," in which Bayes says (in Price's words) that the solution of his problem would not be difficult,

provided some rule could be found according to which we ought to estimate the chance that the probability for the happening of an event perfectly unknown, should lie between any two named degrees of probability, antecedently to any experiments made about it; and that it appeared to him that the rule must suppose the chance the same that it should lie between any two equidifferent degrees; which, if it were allowed, all the rest might be easily calculated in the common method of proceeding in the doctrine of chances. Accordingly, I find among his papers a very ingenious solution to this problem in this way. But afterwards he considered that the *postulate* on which he had argued might not perhaps be looked upon by all as reasonable. (P. 296)

Unfortunately, both Bayes's introduction and his "very ingenious solution" have been lost. But from Price's remarks it is not difficult to guess how the construction would go.

Suppose that a chance setup gives independent and identically distributed (IID) trials with an unknown objective probability for the event in question. If we let X be a variable that counts the number of times that the event in question occurs in n trials, the form of the solution to Bayes's problem is given by

$$\Pr(p_1 \leqslant p \leqslant p_2/X = x) = \frac{\int_{p_1}^{p_2} \binom{n}{x} p^x (1-p)^{n-x} \Pr(dp)}{\int_0^1 \binom{n}{x} p^x (1-p)^{n-x} \Pr(dp)}, \tag{1.6}$$

where $\Pr(*/\dagger)$ is the conditional degree of belief in $*$, given $\dagger$, and $\Pr(dp)$ is the prior distribution over p. To give a definite solution, one needs to specify $\Pr(dp)$. Price remarks that Bayes thought "the rule must be to suppose the chance the same that it [the physical probability p] should lie between any two equidifferent degrees," i.e., $\Pr(dp) = dp$, in which case the solution is given by

$$\Pr(p_1 \leqslant p \leqslant p_2/X = x) = \frac{(n+1)!}{x!(n-x)!} \int_{p_1}^{p_2} p^x (1-p)^{n-x} dp. \tag{1.7}$$

Why must $\Pr(dp)$ be chosen to be uniform? Bayes speaks of an event "concerning the probability of which we absolutely know nothing antecedently to any trials made concerning it" (p. 305). This comment has suggested to some commentators that he thought that the uniformity of the prior was to be justified by an appeal to the principle of insufficient reason. If this were Bayes's reasoning it would be open to the familiar complaints. Thus, for example, Fisher (1922) complained that instead of p we might use another parameter q, where $\sin q = 2p - 1$, and reason that since we antecedently know nothing about q, equal intervals of q ought to be initially regarded as equally probable.

Bayes's actual argument as given in his essay is more subtle and substantial, as first pointed out by Molina (1930, 1931). Bayes gives an operational characterization of an event "concerning the probability of which we absolutely know nothing antecedently"; namely, "concerning such an event I have no reason to think that, in a certain number of trials, it should rather happen any one possible number of times than another" (p. 305). Now *if* $\Pr(\mathrm{d}p) = \mathrm{d}p$, in n trials the marginal distribution is $\Pr(X = x) = 1/(n + 1)$ independently of x, so the operational constraint is satisfied. Conversely, Murray (1930), at Molina's urging, showed that $\Pr(X = x) = 1/(n + 1)$ independently of x for all $n > 0$ *only if* $\Pr(\mathrm{d}p) = \mathrm{d}p$. Thus the operational constraint is equivalent to a uniform prior distribution for p. In modern terms, Bayes was a "predictive Bayesian."[10]

The only thing Molina's account leaves out, as noted by Stigler (1982, 1986), is that the operationalization of the prior distribution is strongly motivated by Bayes's definition of degree of belief as a betting quotient. I would only add that Stigler's insight needs to be pushed further, since the same considerations come into play again when we reach the solution (1.7). For example, in the case where $X = n = 1$, $p_1 = 1/2$, and $p_2 = 1$, (1.7) gives a value of $3/4$. The question now becomes, How do we operationalize $\Pr(1/2 \leqslant p \leqslant 1/X = 1) = 3/4$ in terms of betting behavior, since $1/2 \leqslant p \leqslant 1$ is not an event in the relevant sense?

There is also a second puzzle needing attention. If the Molina and Stigler reconstruction is correct, then Bayes had a persuasive justification for choosing the uniform prior distribution. But then we must ask why he did not simply present this justification straight off. Why did he have to structure his presentation so that the argument from the marginal distribution $\Pr(X = x)$ comes rather late in the essay?

My suggestion will link the two puzzles. Bayes's justification for the uniform prior distribution follows his discussion of the billiard-ball model, and it is this model that provides the proxy event needed to deal with the first puzzle. But before elaborating on this suggestion, I want to examine more fully the notion that Bayes provided a nonproblematic instance of the principle of insufficient reason.

5   Choosing a Prior Distribution

Although ingenious, Bayes's argument for the uniformity of $\Pr(\mathrm{d}p)$ is not as unproblematic as some commentators have suggested (e.g., Molina

1930, 1931; Stigler 1982, 1986). In the first place, the games that can be played with insufficient reason to generate inconsistencies at the level of the unobservable parameter p can also be played at the level of observables. Let n trials of the chance experiment be run, and let Y be a variable whose values index the 2^n possible sequences of successes and failures of the event in question. If complete ignorance about p is to be expressed by equal probabilities for the values of Y and this holds for all $n > 0$, then $\Pr(dp)$ must have all the probability mass concentrated on $p = 1/2$. This example was apparently first posed by Boole (1854, pp. 370–371). It was put forward more recently by Edwards (1978) in a somewhat frivolous vein. The early Wittgenstein would have taken it seriously, since in effect it corresponds to the confirmation function he proposed in the *Tractatus* (see 5.15–5.154). Stigler has complained that the example misses the point of Bayes's strategy, since "X/n estimates the unknown $[p]$, and to have one value of this variable more likely than another is to contradict the hypothesis that we are absolutely ignorant about $[p]$" (1982, p. 256). But this rejoinder seems to be a return to a naked appeal to insufficient reason as applied to p, and as such it is open to Fisher's objection, and it also misses the point of the example. Bayes's strategy was, first, to resist the temptation to say that absolute ignorance about p just means that $\Pr(dp)$ is uniform and, second, to operationalize ignorance about the unobservable p in terms of the distribution of an observable. The point of the objection is that there are different ways to operationalize, and they lead to different $\Pr(dp)$'s.

Of course, Wittgenstein's position can be criticized in various ways. Most obviously, his dogmatic prior distribution prevents any learning from experience: an agent using Wittgenstein's rule will bet at the same odds on the next trial, regardless of the outcomes on the past trials. But whether or not one sees this result as absurd or undesirable depends on one's attitude toward the problem of induction, an aspect of which Bayes was supposed to be solving. If Pr is interpreted as rational or objective degree of belief, as Bayes apparently wanted, then anti-inductivists from Hume through Popper see Wittgenstein's result as an expression of their doctrine that past experience gives no purchase for making justified predictions about the future. Although there is no direct textual evidence one way or the other, it is not too fanciful to suppose that Bayes read Hume's skeptical attacks on induction.[11] If Bayes's essay is to be thought of as a response to Hume, the judgment has to be that Hume was not vanquished.[12]

As an aside it may be worth noting that the connection between Bayes and Hume was solidified by Price. The examples in the appendix, composed by Price, are surely implicit references to Hume. And three years after the appearance of Bayes's essay, Price published his *Four Dissertations* (1767), the fourth of which, entitled "The Importance of Christianity, the Nature of Historical Evidence, and Miracles," used Bayes's work in an attempt to refute Hume's "On Miracles." Upon receiving a copy of Price's book, Hume responded, "I own to you, that the Light, in which you have put this Controversy, is new and plausible and ingenious, and perhaps solid. But I must have more time to weigh it, before I can pronounce this Judgment with Satisfaction to myself."[13] As far as I am aware, Hume did not subsequently directly address Price's arguments.[14]

A second objection to the operationalization of $\Pr(\mathrm{d}p)$ starts from the commonplace that we are rarely in a state of absolute ignorance about the objective probabilities involved in a chance mechanism. Bayes effectively admits as much, since he assumed that the trials are IID, which is to assume quite a lot about the mechanism of a chance experiment. When what we know about the mechanism of the chance experiment favors some values of p over others, what we know may be most naturally expressed directly in terms of $\Pr(\mathrm{d}p)$ rather than in terms of $\Pr(X = x)$. Of course, knowing $\Pr(X = x)$ for all n will determine $\Pr(\mathrm{d}p)$, no matter how lumpy it may be, since $\Pr(X = x)$ will then give all the moments of $\Pr(\mathrm{d}p)$. But I defy any real-world Bayesian agent to specify (without cheating!) enough moments to fix with any tolerable accuracy an even moderately lumpy $\Pr(\mathrm{d}p)$. The orthodox Bayesian personalists will insist that if the enterprise we are engaged in is to make any sense, the agent must be able to assign values to $\Pr(X = x)$ in terms of betting quotients, just as Bayes required. This is correct, but I am suggesting that we make such assignments not by shining some inner light on the distribution of X but by first assessing how the evidence bears on various hypotheses that determine $\Pr(\mathrm{d}p)$ and then calculating the resultant marginal $\Pr(X = x)$. The seeming attractiveness of the operationalization of $\Pr(\mathrm{d}p)$ through the marginal $\Pr(X = x)$ is thus an artifact of the unrealistic assumption of complete ignorance.

It might seem unfair to tax Bayes with such quibbles, since the problem he set himself is limited to events about whose objective probability we know nothing other than that the trials are IID. But Bayes's billiard table, the only concrete example of a chance mechanism offered in his essay, is suggestive of cases that do not conform to this stricture. Imagine a setup (different from Bayes's own, as discussed in the next section) where a single billiard ball is repeatedly tossed on the table, and consider the ball's coming to rest, say, on the right quarter of the table. If, as Bayes suggests, our background knowledge (K) is strongly in favor of the hypothesis of a level table and random throws (H_1), then our $Pr(dp/K)$ will presumably be strongly peaked about $p = 1/4$, and that or something close to it will remain our betting quotient for each of the first few trials, regardless of the outcomes of the preceding trials. Bayes's own rule is incapable of illuminating these intuitions. By contrast, modern Bayesianism can handle this and more complicated situations. Thus, suppose that K is ambiguous, partially favoring H_1 and partially favoring alternative hypotheses, such as that the table is warped in a certain way (H_2), which by itself would lead to a prior strongly peaked about (say) $p = 1/2$. By the principle of total probability,

$$Pr(p_1 \leqslant p \leqslant p_2/(X = x) \& K)$$

$$= \sum_j Pr(p_1 \leqslant p \leqslant p_2/H_j \& (X = x) \& K) \times Pr(H_j/(X = x) \& K), \quad (1.8)$$

where the sum is taken over an exhaustive list of competing hypotheses. Combining what we now call Bayes's theorem,

$$Pr(H_j/E \& K) = \frac{Pr(H_j/K) \times Pr(E/H_j \& K)}{Pr(E/K)}, \quad (1.9)$$

with (1.8), we get, when the new evidence E is $X = x$,

$$Pr(p_1 \leqslant p \leqslant p_2/(X = x) \& K)$$

$$= \frac{\sum_j Pr(p_1 \leqslant p \leqslant p_2/H_j \& (X = x) \& K) \times Pr(H_j/K) \times Pr(X = x/H_j \& K)}{\sum_k Pr(X = x/H_k \& K) \times Pr(H_k/K)}.$$

$$(1.10)$$

Formula (1.10) would not have been regarded by Bayes as a generalization of his solution. Recall that Bayes demanded a rule for calculating numerical values of probabilities. But even if we agree to a rule of equal prior priors for the H_j (i.e., equal values for $Pr(H_j)$), it remains to assign values to the priors $Pr(H_j/K)$. For these latter factors, no mechanical rule suggests itself. As for the general form (1.9) of Bayes's theorem applied, as modern philosophers of science are wont to do, to hypotheses H_j and evidence E of the most varied sort, including nonstatistical theories and nonstatistical experimental outcomes, it is hard to see how the factors involved can be interpreted so as to fit Bayes's framework, either directly as probabilities of events in Bayes's sense or indirectly in terms of proxy events. Bayes is thus prevented by his own scruples from being a Bayesian in the sense that has become popular of late.

6 Proxy Events and the Billiard-Table Model

Postulate 1 of section 2 of Bayes's essay supposes a square billiard table $ABCD$ "to be so made and leveled, that if either of the balls O or W be thrown upon it, there shall be the same probability that it rests upon any one equal part of the plane as another" (p. 302). Ball W is thrown first and a line os is drawn parallel to AD through the point where the center of W comes to rest. Afterward O is thrown n times, and its resting between AD and os constitutes the happening of the event M in a single trial. If we choose units so that the sides of the table have unit length and $0 \leqslant \theta \leqslant 1$ marks the distance between AD and os (see figure 1.1), Bayes's proposition 8 and its corollary correspond respectively to the formulas

$$\Pr((p_1 \leqslant \theta \leqslant p_2) \ \& \ (X = x)) = \int_{p_1}^{p_2} \binom{n}{x} \theta^x (1 - \theta)^{n-x} d\theta \tag{1.11}$$

and

$$\Pr(X = x) = \int_0^1 \binom{n}{x} \theta^x (1 - \theta)^{n-x} d\theta. \tag{1.12}$$

And proposition 9 and its corollary correspond respectively to

$$\Pr(p_1 \leqslant \theta \leqslant p_2 / X = x) = \frac{\int_{p_1}^{p_2} \binom{n}{x} \theta^x (1 - \theta)^{n-x} d\theta}{\int_0^1 \binom{n}{x} \theta^x (1 - \theta)^{n-x} d\theta} \tag{1.13}$$

and

$$\Pr(0 \leqslant \theta \leqslant p_2 / X = x) = \frac{\int_0^{p_2} \binom{n}{x} \theta^x (1 - \theta)^{n-x} d\theta}{\int_0^1 \binom{n}{x} \theta^x (1 - \theta)^{n-x} d\theta} \tag{1.14}$$

In the scholium Bayes notes that the definite integral in (1.12) has the value $1/(n + 1)$, and when we substitute this value in (1.13), we arrive at

$$\Pr(p_1 \leqslant \theta \leqslant p_2 / X = x) = \frac{(n + 1)!}{x!(n - x)!} \int_{p_1}^{p_2} \theta^x (1 - \theta)^{n-x} d\theta. \tag{1.15}$$

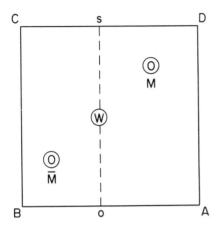

Figure 1.1
Bayes's billiard table

Formula (1.15) has exactly the same mathematical form as the solution formula (1.7) but a much different meaning: (1.15) gives the conditional probability on $X = x$ for the throw of W to have fallen between the distances p_1 and p_2, whereas (1.7) gives the conditional probability for the objective probability p to fall between p_1 and p_2. One can well wonder what role (1.15) has in justifying or motivating (1.7); indeed, one can wonder why the billiard-table model is needed at all. For Bayes's assumption of IID trials and the principles of probability lead directly to (1.6), and Bayes's sophisticated version of the principle of insufficient reason leads from (1.6) to (1.7) without the help of any concrete model or any further assumptions about the chance mechanism producing instances of the unknown event.

The puzzle is resolved by noting that the principles of probability *as Bayes understood them* do not lead directly to (1.6), for as seen above, Bayes's demonstration of these principles depends upon his interpretation of probability as degree of belief in events, and the state $p_1 \leqslant p \leqslant p_2$ is not an event in the relevant sense. By contrast, proposition 8 and its corollary (formulas (1.11) and (1.12)) make good sense for Bayes, since they are stated in terms of the probability of ball W's coming to rest between stated limits. (Proposition 8 reads, "I say that before ball W is thrown, the probability the point o should fall between $[p_1]$ and $[p_2]$ ...," and the corollary reads, "Before the ball W is thrown the probability that the point o will lie somewhere between A and B ... and withal that the event M will happen [x] times. ...") The statement of proposition 9 (formula (1.13)) also begins as a statement about the probability of the resting place of W, but then it shifts to "and consequently that the probability of the event M in a single trial." And the corollary (formula (1.14)) is stated without any reference to the initial throw of W: "if I guess the probability of the event M lies somewhere between 0 and [p_2], my chance to be in the right is. ..." This shift signals Bayes's strategy of having the event $p_1 \leqslant \theta \leqslant p_2$ go proxy for $p_1 \leqslant p \leqslant p_2$.

This strategy is made explicit in the form of a just-so story told in the scholium:

And that the same rule [i.e., one having the same form as (1.15)] is the proper one to be used in the case of an event concerning the probability of which we absolutely know nothing antecedently to any trials made concerning it, seems to appear from the following consideration; viz. that concerning such an event I have no reason to think that, in a certain number of trials, it should rather happen any one possible number of times than another. For, on this account, I may justly reason concerning it *as if* the probability had been at first unfixed, and then determined in such a manner as to give me no reason to think that, in a certain number of trials it should rather happen any one possible number of times than another. But this is exactly the case of the event M. (P. 305; italics added)

The assumption underlying Bayes's problem is that there is a fixed but a priori unknown objective probability for the unknown target event. But, Bayes claims, we can treat the target event as if its probability is at first unfixed and then subsequently determined, as is the objective probability of event M in the billiard model by the rolling of ball W, for the resulting probability of x occurrences of M in n trials is $1/(n+1)$ for every value of x, which is the essential feature of the unknown target event. Thus $p_1 \leqslant \theta \leqslant p_2$ is a suitable proxy for $p_1 \leqslant p \leqslant p_2$, and (1.15) can be used as the solution for the problem.

This reading of Bayes suggests a new interpretation of the worry Price reports in his preface, namely, Bayes's worry was not about the principle of insufficient reason and the choice of a uniform prior distribution per se; indeed, the essay shows no hesitation or hedge on this score. Rather Bayes's worry was how to tie the solution (1.7) to the interpretation of probability he used to justify the probability principles that underlie (1.7). Thus, I suggest that if Bayes had developed a different rule for determining $\Pr(\mathrm{d}p)$ and this rule had given a nonuniform distribution, he would still have needed a proxy event model, even if everyone had agreed that this rule was correct. A warped billiard table, however, would not serve this function—a point brought out by Pearson (1920a, 1920b).

Pearson wanted to make Bayes's billiard table more than merely part of a just-so story. The table, he suggested, could serve to model one mechanism of disease catching if it is assumed that men sicken from a disease whenever their resistance falls below an a priori unknown value $\hat{\theta}$. But with this model it is unnecessary to assume that the playing table is level. For if $\Pr(d\theta) \neq d\theta$, the probability of the occurrence of the disease is $\hat{\theta} = \int_0^{\theta} \Pr(d\theta)$. Thus $\hat{\theta}$ is uniformly distributed, and (1.15) holds with $\hat{\theta}$ in place of θ.[15] Pearson concluded that Bayes was unduly restrictive in assuming the table to be level.

Pearson's conclusion would have merit if Bayes's strategy had been to attack his problem under the assumption that the chance setup involves a mechanism of the billiard-table type. On the contrary, Bayes's posit was that he knew nothing about the chance setup, save that it gives IID trials. His justification for using a uniform prior rests on this posit, and to the extent that this justification succeeds, it does so independently of the billiard-table model, which serves, as I have tried to show, a different function altogether. Moreover, if Bayes had seriously entertained the assumption that as a matter of fact a billiard-table mechanism was operating, then he should have also been prepared to entertain the assumption of other mechanisms that suggest nonuniform priors. But, as argued in section 5, he was not equipped to deal with these latter cases.

A different interpretation of the function of Bayes's billiard-table model is offered by Gillies (1987), who reads Bayes as anticipating and responding to two possible objections that were in fact later raised by Fisher and Neyman. Recall that Fisher's objection was to Bayes's invocation of insufficient reason to justify the uniform prior distribution over p. But here the billiard-table model is irrelevant, since, to repeat, Bayes's justification of the uniform prior distribution, as codified in his scholium, applies to any event "concerning the probability of which we absolutely know nothing antecedently," and it is thus wholly independent of the billiard model. Neyman's (1937) general objection to Bayesianism is based on his contention that, properly speaking, a probability distribution can be assigned only to a random variable, whereas the p Bayes is trying to estimate is a fixed but unknown constant and thus is a random variable in only a degenerate sense at best. I see no plausibility in the notion that Bayes was worried, consciously or unconsciously, about this contention. The probabilities he thought he had to supply for a solution to his problem concern not measures over random variables but (rational?) degrees of belief. Such probabilities do apply to propositions about the values of parameters like p, at least if the propositions can be made to correspond to events in the relevant sense.

7 Applications of Bayes's Result

In the appendix Price discusses what he purports to be applications of Bayes's solution to the problem posed at the beginning of the essay. Dale (1982) has claimed that Price's illustrations are not in fact proper applications of Bayes's solution. This is an issue worth addressing for the light it casts on the meaning and significance of Bayes's accomplishments. For these purposes, it suffices to focus on one of Price's illustrations:

Let us imagine to ourselves the case of a person just brought forth in this world, and left to collect from his observation of the order and course of events what powers and causes take place in it. The Sun would, probably, be the first object that would engage his attention; but after losing it the first night he would be entirely ignorant whether he should ever see it again. He would therefore be in the condition of a person making a first experiment about an event entirely unknown to him. But let him see a second appearance or one *return* of the Sun, and an expectation would be raised in him of a second return, and he might know that there was an odds of 3 to 1 for *some* probability of this. (Pp. 312–313)

If we interpret "*some* probability" to mean that $1/2 \leqslant p \leqslant 1$, then for $X = n = 1$, formula (1.7) gives a value of $3/4$ or, as Price says, odds of 3 to 1.

Dale claims to "see nothing in either the statement of Bayes's problem or its solution (as given in Propositions 9 and 10) that allows Price to solve his problem in the way he does.... It seems clear that the solution is given by Laplace's [value] $2/3$" (1982, p. 44). But from the preceding section we know that Bayes's solution is given not by the propositions of part 2 but by the scholium. Part 2 provides the mathematical form of the solution and allows Bayes to tell his just-so story that links his solution to the interpretation of probability used in part 1. While the propositions of part 2 give the probability that an event has occurred conditional on the assumption that a subsequent event has occurred, the statement of the scholium makes it clear that Bayes intended the solution to apply predictively to future events as well as retrodictively to past events: "In what follows therefore I shall take it for granted that the rule given concerning the event M in proposition 9 is also the rule to be used *in relation to any event* concerning the probability of which nothing at all is known antecedently" (p. 306; italics added). If any further confirmation is needed, it can be found in Bayes's rule 1, which supplies numerical values for the chance (i.e., degree of belief) that the (objective) probability of an event "in a single trial" (p. 308) lies between stated limits. Bayes's scholium, his rule 1, and the very statement of his problem all indicate that he took objective probability to apply to single trials. Thus there is little doubt that Bayes would have regarded Price's illustration as accurate.

Then from whence comes Laplace's value of 2/3? This value is the solution to a different problem. Suppose that after the initial n trials an additional m trials are run. If Z is the variable whose value gives the number of successes in the second run, we can ask for the value of $\Pr(Z = z/X = x)$. Bayes does not give the answer, but it is easily derived from his solution (1.7):

$$\Pr(Z = z/X = x) = \int_0^1 \binom{m}{z} p^z (1 - p)^{m-z} \Pr(dp/X = x)$$

$$= \frac{\binom{m}{z}\binom{n}{x}}{\binom{m+n}{z+x}} \frac{(n+1)}{(n+m+1)} \tag{1.16}$$

When $x = n$ and $z = m$, (1.16) reduces to $(n + 1)/(n + m + 1)$, which is often called Laplace's rule of succession. For Price's sunrise example, $m = n = 1$, which gives $\Pr(Z = 1/X = 1) = 2/3$. In the first part of his article Dale carefully distinguishes the two problems $\Pr(p_1 \leqslant p \leqslant p_2/X = x) = ?$ and $\Pr(Z = z/X = x) = ?$, but in criticizing Price's illustration, he runs them together.[16] There cannot possibly be any conflict between the answers, since the answer to the first entails the answer to the second.

It is Laplace's (1.16) rather than Bayes's (1.7) that is of use in guiding betting behavior. So from the point of view of practical applications, one might say that it is a shortcoming of Bayes's essay that (1.16) is not derived. But this failure only goes to underscore Bayes's goal. Although Bayes uses betting behavior in part 1 to motivate the principles of probability needed for his solution, he was pursuing loftier aims than guiding gamblers; he wanted to provide a reasoned inference to the stable causes or regulations in nature, or an answer to $\Pr(p_1 \leqslant p \leqslant p_2/X = x) = ?$ rather than to $\Pr(Z = z/X = x) = ?.$

8 Conclusion

Philosophers of science with any regard for the history of science must resist the temptation to read Bayes's essay purely from the perspective of contemporary concerns about probability and induction. But unless the interpretations of Bayes that I have offered are well wide of the mark, the attempt to understand Bayes in his own terms reveals a microcosm of foundational problems with which we are still struggling today. In his preface to Bayes's essay, Price laments the then crude state of analogical or inductive reasoning "concerning, which at present, we seem to know little more than that it does sometimes in fact convince us, and at other times not; and that, as it is a means of [a]cquainting us with many truths, of which otherwise we must have been ignorant; so it is, in all probability, the source of many errors, which perhaps might in some measure be avoided, if the force that this sort of reasoning ought to have with us were more distinctly and clearly understood" (p. 297). Price's hope that Bayes's essay would be the key to understanding the force of inductive arguments has in some measure been fulfilled—or so modern Bayesians believe. Whether or not Bayes himself would concur is a matter of intriguing but idle speculation.

It is worth mentioning the irony that what invariably comes to mind when Bayes's name is mentioned, namely "Bayes's theorem," plays no direct role in the solution to his problem. Three ingredients are needed to provide a modern derivation of (1.6). The definition of conditional probability gives

$$\Pr(p_1 \leqslant p \leqslant p_2/X = x) = \frac{\Pr((p_1 \leqslant p \leqslant p_2)\ \&\ (X = x))}{\Pr(X = x)}. \tag{1.17}$$

A form of the principle of total probability is then used to rewrite numerator and denominator of the right-hand side of (1.17) respectively as

$$\Pr((p_1 \leqslant p \leqslant p_2)\ \&\ (X = x))$$

$$= \int_0^1 \Pr_{p'}((p_1 \leqslant p \leqslant p_2)\ \&\ (X = x))\Pr(\mathrm{d}p') \tag{1.18}$$

and

$$\Pr(X = x) = \int_0^1 \Pr_{p'}(X = x)\Pr(\mathrm{d}p'), \tag{1.19}$$

where $\Pr_{p'}(A)$ may be thought of as the probability of A, conditional on the proposition $p = p'$ (see chapter 4). Finally, the application of Lewis's "principal principle" (see chapter 2) allows (1.18) and (1.19) to be rewritten respectively as

$$\Pr((p_1 \leqslant p \leqslant p_2) \,\&\, (X = x)) = \int_{p_1}^{p_2} \binom{n}{x} p'^x (1 - p')^{n-x} \Pr(\mathrm{d}p') \tag{1.20}$$

and

$$\Pr(X = x) = \int_0^1 \binom{n}{x} p'^x (1 - p')^{n-x} \Pr(\mathrm{d}p'). \tag{1.21}$$

What Bayes's theorem and this derivation both bring to the fore is that the prior distribution over p is the key to Bayes's problem. This was Bayes's great insight.

Appendix: Bayes's Definitions and Propositions[17]

Section 1

Definition 1 Several events are *inconsistent*, when if one of them happens, none of the rest can.

Definition 2 Two events are *contrary* when one, or the other of them must; and both together cannot happen.

Definition 3 An event is said to *fail*, when it cannot happen; or, which comes to the same thing, when its contrary has happened.

Definition 4 An event is said to be determined when it has either happened or failed.

Definition 5 The *probability of any event* is the ratio of the value at which an expectation depending on the happening of the event ought to be computed, and the value of the thing expected upon its happening.

Definition 6 By *chance* I mean the same as probability.

Definition 7   Events are independent when the happening of any one of them does neither increase nor abate the probability of the rest.

Prop. 1   When several events are inconsistent the probability of the happening of one or the other of them is the sum of the probabilities of each of them.

Prop. 2   If a person has an expectation depending on the happening of an event, the probability of the event is to the probability of its failure as his loss is if it fails to his gain if it happens.

Prop. 3   The probability that two subsequent events will both happen is a ratio compounded of the probability of the 1st, and the probability of the 2nd on the supposition that the 1st happens.

Cor.   Hence if of two subsequent events the probability of the 1st be a/N, and the probability of both together be P/N, then the probability of the 2nd on the supposition the 1st happens is P/a.

Prop. 4   If there be two subsequent events to be determined every day, and each day the probability of the 2nd is b/N and the probability of both P/N, and I am to receive N if both the events happen the first day on which the second does; I say, according to these conditions, the probability of my obtaining N is P/b.

Cor.   Suppose after the expectation given me in the foregoing proposition, and before it is at all known whether the 1st event has happened or not, I should find that the 2nd event has happened; from hence I can only infer that the event is determined on which my expectation depended, and have no reason to esteem the value of my expectation either greater or less than it was before.

Prop. 5   If there be two subsequent events, the probability of the 2nd b/N and the probability both together P/N, and it being first discovered that the 2nd event has happened, from hence I guess that the 1st event has also happened, the probability I am right is P/b.

Prop. 6   The probability that several independent events shall all happen is a ratio compounded of the probabilities of each.

Prop. 7   If the probability of an event be a, and that of its failure be b in each single trial, the probability of its happening p times, and failing q times in $p + q$ trials is Ea^pb^q if E be the coefficient of the term in which occurs a^pb^q when the binomial $(a + b)^{p+q}$ is expanded.

Section 2

Postulate 1   I suppose that the square table or plane $ABCD$ be so made and levelled, that if either of the balls o or W be thrown upon it, there shall be the same probability that it rests upon any one equal part of the plane as another, and that it must necessarily rest somewhere upon it.

Postulate 2   I suppose that the ball W shall be first thrown, and through the point where it rests a line os shall be drawn parallel to AD, and meeting CD and AB in s and o; and that afterwards the ball O shall be thrown $p + q$ times, and that its resting between AD and os after a single throw be called the happening of the event M in a single trial. These things supposed:

Lem. 1   The probability that the point o will fall between any two points in the line AB is the ratio of the distance between the two points to the whole line AB.

Lem. 2   The ball W having been thrown, and the line os drawn, the probability of the event M in a single trial is the ratio of Ao to AB.

Prop. 8   If upon BA you erect the figure $BghikmA$ whose property is this, that (the base BA being divided into any two parts, as Ab, and Bb and at the point of division b a perpendicular being erected and terminated by the figure in M; and y, x, r representing respectively the ratio of bm, Ab, and Bb to AB, and E being the coefficient of the term in which occurs a^pb^q when the binomial $(a + b)^{p+q}$ is expanded) $y = Ex^py^q$. I say that before the ball W is thrown, the probability that the point o should fall between f and b, any two points named in the line AB, and withall that the event M should happen p times and fail q in $p + q$ trials, is the ratio of $fghikmb$, the part of the figure $BghikmA$ intercepted between the perpendiculars fg, bm raised upon the line AB, to CA the square upon AB.

Cor. (to Prop. 8)   Before the ball W is thrown the probability that the point o will lie somewhere between A and B, or somewhere upon the line AB, and withall that the event M will happen p times and fail q in $p + q$ trials is the ratio of the whole figure AiB to CA. But it is certain that the point o will lie somewhere upon AB. Wherefore, before the ball W is thrown the probability the event M will happen p times and fail q in $p + q$ trials is the ratio of AiB to CA.

Proposition 9   If before anything is discovered concerning the place of the point o, it should appear that the event M has happened p times and failed q in $p + q$ trials, and from hence I guess that the point lies between any two points in the line AB, as f and b, and consequently the probability of the event M in a single trial was somewhere between the ratio of Ab to AB and that of Af to AB; the probability that I am right is the ratio of that part of the figure AiB described as before which is intercepted between perpendiculars erected upon AB at the points f and b, to the whole figure AiB.

Cor. (to Prop. 9)   The same thing supposed, if I guess that the probability of the event M lies somewhere between 0 and the ratio of Ab to AB, my chance to be in the right is the ratio of Abm to AiB.

Scholium

From the preceding proposition it is plain, that in the case of such an event as I there call M, from the number of times it happens and fails in a certain number of trials, without knowing anything more concerning it, one may give a guess whereabouts it's probability is, and, by the usual methods computing the magnitudes of the areas there mentioned, see the chance that the guess is right. And that the same rule is the proper one to be used in the case of an event concerning the probability of which we absolutely know nothing antecedently to any trials made concerning it, seems to appear from the following consideration; viz. that concerning such an event I have no reason to think that, in a certain number of trials, it should rather happen any one possible number of times than another. For, on this account, I may justly reason concerning it as if its probability had been at first unfixed, and then determined in such a manner as to give me no reason to think that, in a certain number of trials, it should rather happen any one possible number of times than another. But this is exactly the case of event M. For before the ball W is thrown, which determines it's probability in a single trial (by cor. prop. 8), the probability it has to happen p times and fail q in $p + q$ or n trials is the ratio of AiB to CA, which ratio is the same when $p + q$ or n is given, whatever number p is; as will appear by computing the magnitude of AiB by the method of fluxions. And consequently before the place of the point o is discovered or the number of times the event M has happened in n trials, I can have no reason to think it should rather happen one possible number of times than another.

In what follows therefore I shall take for granted that the rule given concerning the event M in prop. 9 is also the rule to be used in relation to any event concerning the probability of which nothing at all is known antecedently to any trials made or observed concerning it. And such an event I shall call an unknown event.

2 The Machinery of Modern Bayesianism

I. J. Good once quipped that there are more forms of Bayesianism than there are actual Bayesians. While the ever growing popularity of Bayesianism may have invalidated the letter of this quip, its core message is still sound: there are many rooms to the mansion that Bayes helped to build. No attempt will be made here to systematically survey all of this real estate. Bayesians of whatever persuasion can speak for themselves; indeed, they do speak for themselves, often *ad nauseam*. My focus will be kept on issues concerning the testing and confirmation of scientific hypotheses and theories, typically of a nonstatistical kind. Bayesian personalism will be the starting point for most of my investigations. Issues in Bayesian decision theory and technical issues in Bayesian statistics will be largely ignored, although from time to time technicalia will intrude.

1 The Elements of Modern Bayesianism

Bayesians of all stripes are united in the convictions that qualitative approaches to confirmation, such as hypotheticodeductivism and Hempel's instance confirmation (see chapter 3), are hopeless and that an adequate accounting of the way evidence bears on hypotheses and theories must be quantitative. The form of Bayesianism I will track here follows in Thomas Bayes's footsteps by implementing the quantitative approach in terms of degrees of belief regimented according to the principles of the probability calculus. The form of probability theory needed for applications to issues of confirmation will be presented in section 2. Bayes, as we saw in chapter 1, exploited the connection between degrees of belief and betting behavior in an attempt to justify the principles of probability. Modern Bayesians follow suit with their Dutch-book arguments, which will be examined in sections 3 and 4.

Bayesians are also united on the importance of Bayes's theorem, a result that Bayes himself never stated in modern form. If H, K, and E are respectively the hypothesis at issue, the background knowledge, and the new evidence, then one form of Bayes's theorem states that

$$\Pr(H/E \ \& \ K) = \frac{\Pr(H/K) \times \Pr(E/H \ \& \ K)}{\Pr(E/K)}. \tag{2.1}$$

If $\{H_i\}$, $i = 1, 2, \ldots$, is a set of mutually exclusive and exhaustive hypotheses, the principle of total probability allows (2.1) to be rewritten as

$$\Pr(H_i/E \ \& \ K) = \frac{\Pr(H_i/K) \times \Pr(E/H_i \ \& \ K)}{\sum_j \Pr(E/H_j \ \& \ K) \times \Pr(H_j/K)}. \tag{2.2}$$

In Bayesian accounts of confirmation, the explanations of confirmational virtues are couched largely in terms of the factors on the right hand sides of (2.1) and (2.2): $\Pr(H/K)$, the *prior probability* of H; $\Pr(E/H \ \& \ K)$, the *likelihood* of E on H and K; and $\Pr(E/K)$, the *prior likelihood* of E.

The forms of Bayesianism to be examined here also share the tenet that learning from experience is to be modeled as conditionalization. The rule of *strict conditionalization* says that if it is learned for sure that E and if E is the strongest such proposition, then the probability functions $\Pr_{old}$ and $\Pr_{new}$, representing respectively degrees of belief prior to and after acquisition of the new knowledge, are related by

$$\Pr_{new}(\cdot) = \Pr_{old}(\cdot/E). \tag{SC}$$

Bayes's proposition 5 can, as we saw in chapter 1, be regarded as an attempt to justify this rule. From the point of view of strict conditionalization, Bayes's theorem (2.1) makes explicit how the acquisition of new evidence impacts on previous degrees of belief to produce new degrees of belief.[1]

A more sophisticated form of conditionalization that allows for uncertain learning is due to Richard Jeffrey (1983b). If we observe a jelly bean by dim and flickering candle light, we will rarely come away with certain knowledge of the color of the bean, but our probabilities will have changed.[2] We may have gone, for example, from complete ignorance as to whether the bean is red, yellow, or green (as represented by probabilities of 1/3 for each) to, say, a probability of 2/3 for red and a probability of 1/6 each for yellow and green. To generalize and formalize, let $\{E_i\}$, $i = 1, 2, \ldots$, be a partition of the probability space. Intuitively, the belief change that takes place is supposed to be generated by the way in which the experience bears on this partition (e.g., the color partition in the above example). The belief change then accords with *Jeffrey conditionalization* just in case

$$\Pr_{new}(A) = \sum_i \Pr_{old}(A/E_i) \times \Pr_{new}(E_i) \quad \text{for all } A. \tag{JC}$$

Strict conditionalization is the special case where the new probability of one of the elements of the partition is one. An application of total probability shows that (JC) obtains under the condition of *rigidity*:

$Pr_{new}(A/E_i) = Pr_{old}(A/E_i)$ for all A and all i (R)

Arguably, (R) should apply in the jelly bean case when we look but don't get to touch, smell, or taste the bean, so that any change in our degrees of belief about the sweetness, scent, or texture of the bean should be due entirely to changes in our degrees of belief about its color.[3]

In chapter 1 we saw that Bayes's essay contained a tension between personalism (probability as personal degree of belief) and objectivism (probability as uniquely determined rational degree of belief). The tension survives in modern Bayesianism. The *pure personalists*, as represented by de Finetti and his followers, recognize the axioms of probability as the only synchronic constraints on degrees of belief. Some personalists have also refused to recognize any diachronic constraints, but it turns out that the Dutch-book arguments used to justify the probability axioms can also be used to justify rules of conditionalization (see, however, section 6 below). *Tempered personalists* would add further constraints, such as Lewis's principal principle to be discussed below in section 7, or Shimony's (1970) injunction on the members of a scientific community to assign a nonzero prior to any hypothesis seriously proposed by a fellow member of the community. *Objectivists*, such as Harold Jeffreys (1961, 1973), carry the tempering of priors to the extreme by proposing principles to uniquely fix these numbers. Thomas Bayes himself seems to have fallen into this camp, at least with respect to the problem treated in his founding essay.

The implications of these differing forms of Bayesianism for confirmation theory will be discussed in later chapters. The present chapter concentrates on an elementary exposition of the common core of all forms of Bayesian personalism.

2   The Probability Axioms

Since propositions are the object of belief and since probability is being interpreted as degree of belief, probabilities will be assigned to objects that express propositions, namely sentences. More specifically, let $\mathscr{A}$ be a collection of sentences. The content and structure of $\mathscr{A}$ will vary from context to context, but at a minimum it is assumed that $\mathscr{A}$ is closed under finite truth-functional combinations. Then a *probability function* Pr is a map from $\mathscr{A}$ to $\mathbb{R}$ satisfying at least the following restrictions:

$\text{Pr}(A) \geqslant 0 \quad \text{for any } A \in \mathscr{A}$ \hfill (A1)

$\text{Pr}(A) = 1 \quad \text{if} \models A$ \hfill (A2)

$\text{Pr}(A \lor B) = \text{Pr}(A) + \text{Pr}(B) \quad \text{if} \models \neg(A \& B)$ \hfill (A3)

Here $\models A$ means that A is valid in the sense that A is true in all models or all possible worlds.[4] Again, the content and structure of the models or possible worlds will depend upon the context. I assume at a minimum that $\mathscr{A}$ respects propositional logic.[5] In this case (A1) to (A3) suffice to prove many of the familiar principles of probability, including the following:

$\text{Pr}(\neg A) = 1 - \text{Pr}(A)$ \hfill (P1)

$\text{Pr}(A) = \text{Pr}(B) \quad \text{if} \models A \leftrightarrow B$ \hfill (P2)

$\text{Pr}(A \lor B) = \text{Pr}(A) + \text{Pr}(B) - \text{Pr}(A \& B)$ \hfill (P3)

$\text{Pr}(A) \leqslant \text{Pr}(B) \quad \text{if} A \models B$ \hfill (P4)

Here $A \models B$ means that A semantically implies B in the sense that B is true in every model or possible world in which A is true.

Conditional probability may be introduced as a defined concept:

Definition   If $\text{Pr}(B) \neq 0$, then $\text{Pr}(A/B) \equiv \text{Pr}(A \& B)/\text{Pr}(B)$.

Bayes's theorem is now a simple consequence of this definition. An alternative approach takes conditional probability $\text{Pr}(\cdot/\cdot)$ as primitive and defines the associated unconditional probability $\text{Pr}(\cdot)$ as $\text{Pr}(\cdot/N)$, where N is a necessary truth (i.e., $\models N$). The advantage of this approach is that $\text{Pr}(A/B)$ can be defined even when $\text{Pr}(B) = 0$. Conditional probability is discussed in more detail in appendix 1 to this chapter.

Some of the applications to be considered in later chapters also assume a principle of continuity.

C   If $A_i \in \mathscr{A}$, $i = 1, 2, \ldots$, are such that $A_{n+1} \models A_n$ for each n and $\{A_1, A_2, \ldots\}$ is inconsistent (i.e., the A_i are not all true in any model or possible world), then $\lim_{n \to \infty} \text{Pr}(A_n) = 0$.

Actually, the axiom I will use most often is a weaker principle that applies to first-order predicate logic. Let 'P' be a monadic predicate and let $a_1, a_2, \ldots$ be a countably infinite sequence of individual constants. The principle added as an additional axiom asserts that

$$\Pr((\forall i)Pa_i) = \lim_{n \to \infty} \Pr\left(\underset{i \leqslant n}{\&} Pa_i\right), \tag{A4}$$

where $\underset{i \leqslant n}{\&} Pa_i$ stands for $Pa_1 \,\&\, Pa_2 \,\&\, \dots \,\&\, Pa_n$. If we require that $(\forall i)Pa_i \models Pa_n$ for every n and that $\{\neg(\forall i)Pa_i, Pa_1, Pa_2, \dots\}$ be inconsistent, then (A4) is shown to be a consequence of (C) by one's taking $A_n \equiv (\underset{i \leqslant n}{\&} Pa_i \,\&\, \neg(\forall i)Pa_i)$. It also follows that $\Pr((\exists i)Pa_i) = \lim_{n \to \infty} \Pr(\underset{i \leqslant n}{\bigvee} Pa_i)$, where $\underset{i \leqslant n}{\bigvee} Pa_i$ stands for $Pa_1 \lor Pa_2 \lor \dots \lor Pa_n$. Axiom (A4) can be regarded as an extension of the finite additivity principles (A3) and (P3) to countable additivity.

In a manner of speaking, "half" of (A4) is already a consequence of (A1) through (A3). Since $(\forall i)Pa_i \models \underset{i \leqslant n}{\&} Pa_i$, it follows by (P4) that $\Pr((\forall i)Pa_i) \leqslant \Pr(\underset{i \leqslant n}{\&} Pa_i)$. Moreover, $\Pr(Pa_1), \Pr(Pa_1 \,\&\, Pa_2), \Pr(Pa_1 \,\&\, Pa_2 \,\&\, Pa_3), \dots$ is a monotone decreasing sequence bounded from below (by (A1)), and so it must have a limit. Thus $\Pr((\forall i)Pa_i) \leqslant \lim_{n \to \infty} \Pr(\underset{i \leqslant n}{\&} Pa_i)$. To turn the '$\leqslant$', into an '$=$', as required by (A4), requires a new substantive assumption.

Continuity or countable additivity does not come without intuitive cost. Consider a denumerably infinite list $H_1, H_2, \dots$ of pairwise incompatible and mutually exhaustive hypotheses. One might think that it should at least be possible to treat these hypotheses in an evenhanded manner by assigning them all the same probability. But this we cannot do consistently with (C), since (C) implies that $\sum_{i=1}^{\infty} \Pr(H_i) = 1$. Continuity thus forces us to play favorites. (Sticking to finite additivity would allow for a draconian evenhandedness in the form $\Pr(H_i) = 0$ for all i.) On the other hand, abandoning countable additivity leads to results that Bayesians and non-Bayesians alike find repugnant. Some of these results will be discussed in appendix 1.

A different nomenclature is presupposed when mathematicians and statisticians speak of probability. For them, a probability space is a triple $(\Omega, \mathscr{F}, \mathscr{P})$. Ω, a set of elements, is called the sample space; $\mathscr{F}$, a field of subsets of Ω, is the collection of measurable sets; and $\mathscr{P}$ is a nonnegative (finitely or countably additive) function from $\mathscr{F}$ to $\mathbb{R}$. (Here countable additivity means that if $B_i \in \mathscr{F}$, $i = 1, 2, \dots$ are pairwise disjoint, then $\mathscr{P}(\bigcup_{i=1}^{\infty} B_i) = \sum_{i=1}^{\infty} \mathscr{P}(B_i)$.) As is discussed in detail in chapter 6, one can move from the Bayesian personalist conception of probability to the mathematical conception by taking Ω to be the set of models of the language of $\mathscr{A}$, $\mathscr{F}$ to be a field generated by sets of models of the form $\mathrm{mod}(A)$ for a sentence $A \in \mathscr{A}$, and $\mathscr{P}_{\iota}$ to be a measure satisfying $\mathscr{P}_{\iota}(A) = \mathscr{P}_{\iota}(\mathrm{mod}(A))$. One can also move in the opposite direction, although an awkwardness occurs when $\mathscr{F}$ is a σ field (see appendix 2) and probabilities qua degrees of belief are assigned to sentences in a standard first-order language, for then not every member of $\mathscr{F}$ will correspond to a sentence, since these languages do not allow infinite conjunctions or disjunctions. We can often smooth over this awkwardness by taking limits of probabilities of finite conjunctions or disjunctions.

3   Dutch Book and the Axioms of Probability

Rather than simply assuming that degrees of belief are regimented by the principles of probability, one could try to exploit the interpretation of probability as degree of belief as a means of getting a justification for the probability axioms. We saw in chapter 1 that Thomas Bayes took this tack by using the connection between degrees of belief and betting behavior. Ramsey (1931) and de Finetti (1937) followed a related tack with their Dutch-book strategy, although they were apparently unaware of the details of Bayes's work, which contains, as we have seen in chapter 1, intimations of Dutch book. The presentation given here follows Shimony 1955.

By a *bet* on $A \in \mathscr{A}$ let us understand a contractual arrangement between a bettor and a bookie by which the bettor agrees to pay the bookie the amount \$b if A turns out to be false and the bookie agrees to pay the bettor \$a if A turns out to be true. The sum \$(a + b)$ is called the *stakes* of the bet, and the ratio b/a is called the bettor's *odds*. If Pr is the bettor's degree-of-belief function, the expected monetary value of the bet for him is \$a \times Pr(A) - \$b \times Pr(\neg A)$. The bet is said to be *fair* (respectively, *favorable*, *unfavorable*) to the bettor according as the expected value is zero (respectively, positive, negative). Using the negation principle (P1), the condition for a fair bet comes to $Pr(A) = b/(a + b)$. This ratio is called the bettor's *fair betting quotient*.

The idea of the Dutch-book argument is to turn this construction around to produce a justification of the probability axioms: assume that degree of belief functions as a fair betting quotient and then show that something very nasty will happen if the degrees of belief fail to conform to the probability axioms. Thus if $Pr(A) = r$ is your degree of belief in A, then (the story goes) you should be willing to bet on A on the terms in table 2.1. S is allowed to be either positive or negative, which means that you are

Table 2.1
Terms for betting on A

	Pay	Collect	Net
A false	rS	0	$-rS$
A true	rS	S	$(1-r)S$

Note: S stands for the stakes.

required to accept either end of the bet. If you do enter such an arrangement, the nasty thing that threatens is Dutch book, a finite series of bets such that no matter what happens, your net is negative (a violation of what is called *coherence* for degrees of belief). The *Dutch-book theorem* shows that if any one of the axioms (A1) to (A3) is violated, then Dutch book can be made. The *converse Dutch-book theorem* shows that if (A1) through (A3) are satisfied, then Dutch book cannot be made in a finite series of bets. This converse is crucial to the motivation for conforming degrees of belief to the principles of probability, for if such a conformity were not guarantee against Dutch book, the threat of Dutch book would not be a very effective inducement to conformity. Only the proof of the Dutch-book theorem will be sketched here. The interested reader can consult Kemeny 1955 and Lehman 1955 for the converse.

To establish that (A1) is necessary to avoid Dutch book, suppose that $\Pr(A) = r < 0$. Choose $S < 0$ and note that the net is negative whether or not A is true. Similarly, if $\Pr(A) = r > 1$, choosing $S > 0$ leads to a loss, come what may. We can now establish that (A2) is necessary to avoid Dutch book. For suppose that $\Pr(A) = r \neq 1$ even though $\models A$. By the previous results, $0 \leqslant r \leqslant 1$. Choosing $S < 0$ then leads to a loss in case A is true, which is the only possible case. Finally, to show the necessity of (A3), suppose that $\models \neg(A \,\&\, B)$ and consider a series of three bets: one on A with a betting quotient $\Pr(A) = r_1$ at stakes S_1, one on B with a betting quotient $\Pr(B) = r_2$ at stakes S_2, and one on $A \lor B$ with a betting quotient $\Pr(A \lor B) = r_3$ at stakes S_3. There are three possible cases to consider (table 2.2). The theory of linear equations then shows that the stakes can be chosen so that the nets are all negative unless $r_3 = r_1 + r_2$, i.e., unless (A3) holds.

If regarded as a definition, the formula given in section 1 for conditional probability does not stand in need of a justification. But as in de Finetti 1937, the notion of the conditional probability of B on A can be introduced

Table 2.2
Net payoffs for the three bets taken together

	Net
A true, B false	$(1 - r_1)S_1 - r_2 S_2 + (1 - r_3)S_3$
A false, B true	$-r_1 S_1 + (1 - r_2)S_2 + (1 - r_3)S_3$
A false, B false	$-r_1 S_1 - r_2 S_2 - r_3 S_3$

as a primitive and then operationalized in terms of a bet on B conditional on A, the terms of which specify that if A obtains, a standard unconditional bet on B is in effect, whereas if A fails, the bet is called off. Then (the story goes) $Pr(B/A)$ should be the agent's critical odds for this conditional bet. The agent is now offered three bets: a standard bet on A, a standard bet on $B \& A$, and a bet on B conditional on A. It is left as an exercise to show that unless $Pr(B/A) \times Pr(A) = Pr(B \& A)$, stakes can be chosen for the three bets so that the agent has a sure net loss. This argument does *not* justify the rule of conditionalization, which requires a different argument (see section 5 below).

The Dutch-book justification for continuity is not so pretty, and this is perhaps one of the reasons it plays no role in the Bayesianism of Ramsey, de Finetti, and Savage.[6] To Dutch-book a violation of (C) or (A4), which is not also a violation of (A1) through (A3), requires laying an infinite series of bets. But if I were to risk the same finite amount, no matter how small, on each of these bets, then I would have to have an infinite bankroll, an impossible dream. And if the dream should come true, I would not care one whit about losing a finite or even an infinite sum if, as can always be arranged, I have an infinite amount left over. To remedy this defect, we can imagine that the bettor accepts an infinite series of fair bets but that the total amount he risks is finite; e.g., he risks $\$(1/2)$ on the first bet, $\$(1/4)$ on the second, $\$(1/8)$ on the third, and so on. With this setup, Adams (1961) shows that a sure loss results from a violation of the general continuity axiom (C) (see also Spielman 1977).

4   Difficulties with the Dutch-Book Argument

Qualms about the Dutch-book justification of the probability axioms are so numerous and diverse that it is hard to classify them. For future reference I note that when the requirement of logical omniscience is dropped, as it must be for realistic agents, the situation becomes more complicated; this matter is discussed in chapter 5. For the present context, which takes logical omniscience for granted, I begin with three miscellaneous qualms. First, the Dutch-book construction for countable additivity involves, in Ernest Adams's words, "extremely unrealistic systems" (1961, p. 8). For those who insist that degrees of belief must be operationalized in terms of economic transactions, this constitutes a reason to reject countable additivity. (Thus it is not surprising that countable additivity plays no role in de Finetti's personalism.) But for those of us who reject operationalism and behaviorism and insist that countable additivity is needed, the difficulty is a shortcoming of the Dutch-book construction. Second, the requirement that the agent be willing to take either side of the bet (i.e., the stakes S may be either positive or negative) may not be satisfied by actual gamblers, and in any case it already assumes the negation principle.[7] Third, a Bayesianism that appeals to both Dutch book and strict conditionalization is on a collision course with itself. The use of strict conditionalization leads to situations where $Pr(A) = 1$ although $\not\models A$. As a result, something almost as bad as Dutch book befalls the conditionalizer; namely, she is committed to betting on the contingent proposition A at maximal odds, which means that in no possible outcome can she have a positive gain and in some possible outcome she has a loss (a violation of what is called *strict coherence*). It is too facile to say in response that this is a good reason for abandoning strict conditionalization in favor of Jeffrey conditionalization or some other rule for belief change; for all the results about merger of opinion and convergence to certainty so highly touted in the Bayesian literature depend on strict conditionalization (see chapter 6).

A more basic worry harkens back to Bayes's insistence that probability as a betting quotient be attached to "events," i.e., decidable propositions (see chapter 1). Bets on the outcome of the Kentucky Derby are one thing, bets on scientific hypotheses are quite another. A hypothesis with the quantifier structure $(\exists x)(\forall y)Rxy$ can be neither verified nor falsified by finite means. Thus a bet on such a hypothesis turns on a contingency that can never be known for certainty to hold or to fail, and so the parties to the bet have no sure way to settle the matter. To try to settle the bet by appeal to the probable truth or falsity of the hypothesis runs afoul of the fact that the parties can and often do disagree on whether the hypothesis is probably true. But if the bet is never paid off, fear of being bilked disappears.

The response to this worry might be that bookies wearing wooden shoes, money pumps, etc. are just window dressing. The underlying assumption is that degrees of belief are manifested in preferences over the kinds of bets described in section 3. This assumption granted, the Dutch-book construction stripped of its decoration shows that the failure of degrees of belief to conform to the probability calculus results in a structural inconsistency in the individual's preferences. Suppose that the individual is nonsatiated in that she prefers more money to less. Then if this person violates (A1) or (A2), the Dutch-book construction reveals that she is literally inconsistent with herself, since she prefers the certainty of handing over some $\$\varepsilon > 0$ to the status quo, despite her professed nonsatiation. In the case of (A3) the argument is more involved, since it appeals to another principle, "the package principle"; to wit, a person's preferences are inconsistent if there is a finite series of bets such that she regards each as preferable to the status quo while at the same time she regards the status quo as preferable to the package of bets. If this hypothetical agent violates (A3), we proceed to construct a finite series of bets each of which she finds favorable. By the package principle, she should then find the package favorable. But the package is shown to be equivalent to handing over $\$\varepsilon > 0$, which contradicts nonsatiation. Note that on this reading the Dutch-book construction does not justify strict coherence, i.e., the requirement that $Pr(A) = 1$ only if $\models A$, which I take to be a mark in favor of this reading.

Schick (1986) has questioned the normative status of the package principle. Its plausibility, he argues, rests on accepting the notion of value additivity, which holds that the value of the package of bets is the sum of the values of the individual bets. But, Schick claims, an agent who refuses to conform her degrees of belief to the probability axioms may read the Dutch-book construction as a reason to reject value additivity. Schick's objection may not at first seem very moving, but it gains force in the context of the sequential decision making that comes into play in the attempted diachronic Dutch-book justification for conditionalization (see section 6).

Although the above reconstrual of the Dutch-book construction is a step forward, it is still too closely tied to the behavioristic identification of belief with dispositions to place bets. Once it is admitted that betting behavior is only indicative of, and not constitutive of, underlying belief states, it must also be admitted that belief and behavior are mediated by many factors and that these factors can weaken to the breaking point the simpleminded linkage assumed in the Dutch-book construction. In poker, for example, betting high may be a good way to scare off the other players and win the pot (see Borel 1924). And generally, a knowledge of the tendencies of opponents may make it advisable to post odds that differ from one's true probabilities (see Adams and Rosenkrantz 1980).[8]

Two responses can be made to this complaint. First, one can drop the Dutch-book approach in favor of a justification of the probability axioms that focuses directly on the nature of belief and the cognitive aims of inquiry and eschews altogether preferences for goodies, monetary or otherwise. Some candidates for such a justification will be examined in the next section. Second, one can continue to push the Dutch-book approach by taking into account in a more systematic manner the preference structure of the agent. I will follow this theme in the remainder of this section.

The opening melody of this theme is that the Dutch-book construction rests on the assumption that utility is linear with money, or equivalently, that agents are risk neutral, an assumption known to be false for many if not most real-world agents.[9] To illustrate the complications that can arise in trying to use betting behavior to elicit degrees of belief for such real-world agents, let us analyze from the point of view of expected-utility theory the elicitation device Bayes himself used. Let q be the maximum amount the agent is willing to pay for a contract that awards r if A is true and $0 otherwise. If U is the agent's utility function and $\Pr(dw/A)$ and $\Pr(dw/\neg A)$ are the agent's conditional probability distributions for wealth exclusive of the contract prize, then a little algebra shows that the expected-utility hypothesis implies that the agent's degree of belief in A is

$$1 \bigg/ \left[1 + \frac{\int (U(w + r - q) - U(w)) \Pr(dw/A)}{\int (U(w) - U(w - q)) \Pr(dw/\neg A)} \right]$$

(see Kadane and Winkler 1987). If the agent is risk neutral, i.e., if U is linear, then the degree of belief is seen to be equal to q/r, as Bayes thought. If $\Pr(dw/A) = \Pr(dw/\neg A)$ (i.e., the agent's wealth apart from the contract payoff is not probabilistically dependent on A) but the agent is not risk neutral, then $\Pr(A)$ will differ from q/r: if the agent is risk-averse, q/r will understate $\Pr(A)$, while if she is risk-positive, q/r will overstate $\Pr(A)$. And if $\Pr(dw/A) \neq \Pr(dw/\neg A)$, q/r is an even more distorted measure of $\Pr(A)$.

The moral is that the direct elicitation of degrees of belief by betting behavior is doomed to failure. Degrees of belief and utilities have to be elicited in concert. In the standard developments of this concerted elicitation the aim is to show that preferences satisfying (what are taken to be) rationality constraints can be represented in terms of expected utility, with the probabilities being uniquely determined and the utilities determined up to positive linear transformations. But the alleged rationality constraints are open to challenge (see, for example, the paradoxes in Allais 1953 and Ellsberg 1961). Moreover, when the utilities are dependent not just on the prizes but also on the propositions whose utilities are being elicited, then the probabilities may not be uniquely determined (see Schervish, Seidenfeld, and Kadane 1990 and Seidenfeld, Schervish, and Kadane 1990). Here I must break off the discussion, since I have strayed beyond the scope of this work.

5   Non-Dutch-Book Justifications of the Probability Axioms

Aside from a fear of being bilked by Dutch bookies, there are a number of other motivations for conforming degrees of belief to the probability calculus, three of which will be mentioned here.

The first is articulated by Rosenkrantz (1981), who follows de Finetti (1972). Consider a partition $\{H_i\}$, $i = 1, 2, \ldots, N$, and an agent who distributes her degrees of belief x_i over the H_i in accord with the constraint that $0 \leqslant x_i \leqslant 1$ but not necessarily obeying the condition $\sum_i x_i = 1$, as would be the case if she obeyed the probability calculus. Suppose that when H_j is the true hypothesis, the inaccuracy of her degrees of belief is measured by the least-squares function

$$I(x; H_j) \equiv x_1^2 + \cdots + x_{j-1}^2 + (1 - x_j)^2 + x_{j+1}^2 + \cdots + x_N^2 \qquad (2.3)$$

If the x_i do not sum to 1, there is an alternative set of degrees of belief y_i that do sum to 1 and that dominate the x_i in the sense that $I(y; H_j) < I(x; H_j)$, whatever the value of j. This conclusion continues to hold when (2.3) is generalized to a weighted least-squares measure where the weights reflect judgments of how far the false alternatives are from the true hypothesis. If the conclusion could be further generalized to any "reasonable" measure of inaccuracy, we would be entitled to draw the moral that failure to obey the axioms of probability undermines the goal of accuracy. A discussion of what conditions constitute a reasonable measure of inaccuracy, together with a review of results and conjectures about the sought after generalization, are found in Rosenkrantz 1981 (see also Lindley 1982).

A second kind of justification is best construed as directed at well-tempered personalists who aim at rational degrees of belief. It can be found in various versions in Aczél 1966; Cox 1946, 1961; Good 1950; and also in Shimony 1970, the version I will report here. It works on the concept of conditional probability. Let $\Pr(H/E)$ be a real-valued function defined on pairs of sentences (H, E), where H is a member of a nonempty set $\mathscr{A}$ of sentences closed under truth functional operations and E is a member of the noncontradictory elements $\mathscr{B}^0$ of $\mathscr{B} \subseteq \mathscr{A}$ (see appendix 1). It is further supposed that $\Pr(\cdot/\cdot)$ satisfies the following six conditions:

C1 $\Pr(H/E) = \Pr(H'/E')$ if $\models H \leftrightarrow H'$ and $\models E \leftrightarrow E'$.

C2 For any $E \in \mathscr{B}^0$, there is an r_0 such that for any contradiction $C \in \mathscr{A}$ and any $H \in \mathscr{A}$, $\Pr(C/E) = r_0 \leqslant \Pr(H/E)$.

C3 There is an r_1 such that for all $E, F \in \mathscr{B}^0$, $\Pr(E/E) = \Pr(F/F) = r_1 > r_0$.

C4 $\Pr(H \,\&\, E/E) = \Pr(H/E)$.

C5 For any $E, F \in \mathscr{B}^0$, there is a function f_E such that $\Pr(H \,\&\, F/E) = f_E(\Pr(H/F \,\&\, E), \Pr(F/E))$.

C6 For any $E \in \mathscr{B}^0$, there is a continuous and monotone increasing function g_E in both variables such that if $E \models \neg(H \,\&\, J)$, then $\Pr(H \vee J/E) = g_E(\Pr(H/E), \Pr(J/E))$.

Then there exists a continuous and monotone increasing function h such that $h(r_0) = 0$, $h(r_1) = 1$, and $\widehat{\Pr}(H/E) = h(\Pr(H/E))$ satisfies the standard axioms for conditional probability.

The usefulness of this technical result for the justification of the probability axioms depends on the persuasiveness of two further assumptions: first, that (C1) through (C6) should be satisfied for any rational conditional degree of belief function and, second, that if Pr is a suitable measure of rational degree of belief, then so is any monotone function of Pr, which leaves us free to choose a Pr that satisfies the standard axioms. Neither of these assumptions recommends itself with overwhelming force.

A third mode of justification starts from Carnap's (1950) remark that rational degrees of belief can, in some instances, be construed as estimates of relative frequencies. Thus if H is of the form 'Pa', my degree of belief in H may be interpreted as my estimate of the relative frequency of individuals with the property designated by 'P' in some appropriate reference class.[10] If my personal probabilities for propositions of this form are not to be precluded a priori from being accurate estimates of frequencies, they must fulfill the standard probability axioms, since frequencies do (see van Fraassen 1983a and Shimony 1988).

Although attractively straightforward, such frequency-driven justifications have their limitations. As a result of calculation or of consulting theories like quantum mechanics, my degree of belief in H may be an irrational number. If 'frequency' means finite frequency, i.e., the ratio of the number of individuals that have the property to the total number of individuals in the (finite) reference class, then I am automatically precluded from having an exactly accurate estimate. Limiting relative frequencies in infinite sequences do not share this shortcoming, but such frequencies can lead via the continuity axiom to a conflict with other probability assignments we may want to make. Thus, for example, my estimate of the limiting relative frequency for events such as Pa_i may be 0 for each i, in which case I set $\Pr(\bigvee_{i \leqslant n} Pa_i) = 0$ for every n. But at the same time I may be convinced that at least one of the individuals must be a 'P', which contradicts (A4). More generally, for the multiply quantified hypotheses encountered in the advanced sciences, there is no obvious or natural way in which one's degree of belief can be regarded as an estimate of relative frequency in either the finite or limiting sense. Of course, I can calibrate my degree of belief in H with frequencies by finding an H' such that $\Pr(H) = \Pr(H')$ and such that $\Pr(H')$ does have a natural interpretation as an estimate of a frequency. But without further restrictions, there is no guarantee that the probabilities assigned to the class of hypotheses so calibrated will satisfy the probability axioms.

Although Dutch book and the other methods of justification investigated in this section are all subject to limitations and objections, collectively they provide powerful persuasion for conforming degrees of belief to the probability calculus.

6 Justifications for Conditionalization

Dutch-book justifications can be given for both strict conditionalization (Teller 1973, 1976) and Jeffrey conditionalization (Skyrms 1987).[11] To consider the former, suppose without any real loss of generality that upon learning E the agent shifts from Pr_{old} to Pr_{new}, where $y = \text{Pr}_{old}(A/E) - \text{Pr}_{new}(A) > 0$ and $x = \text{Pr}_{old}(A/E) > 0$. The diachronic Dutch bookie first sells the agent three bets b_1: [$1; A \& E$], b_2: [$x; \neg E$], and b_3: [$y; E$], at what the agent computes to be their fair values. (Recall that [$z; C$] stands for the contract that pays z if C obtains and 0 otherwise.) If E proves to be false, the agent has a net loss of $y\text{Pr}_{old}(E)$. On the other hand, if E turns out to be true, the bookie buys back from the agent the bet b_4: [$1; A$] for its then expected value to the agent ($\text{Pr}_{new}(A) = (\text{Pr}_{old}(A/E) - y)$). The agent then has a net loss of $y\text{Pr}_{old}(E)$, regardless of whether A obtains.

We can assess this argument for conditionalization in the light of the distinction drawn above in section 4 between two readings of the Dutch-book construction. If the central concern is to escape being systematically bilked by a bookie, there is a simple solution that doesn't commit you to conditionalization: don't publicly announce your strategy for changing belief in the face of new evidence. If you are worried about clairvoyant bookies who can read your mind, then don't make up your mind in advance; just wait to see what evidence comes in and then wing it. (This is, in fact, what many of us do in practice.) This will make you proof against systematic bilking, save by those bookies who have the ability to foresee your future belief states. But from such precognitive bookies not even good Bayesian conditionalizers are safe. Of course, if you do not conditionalize, there will be a hypothetical lucky bookie who by chance rather than system hits on a series of bets that guarantees you a net loss, but then even if you do conditionalize, there will be a hypothetical lucky bookie who takes you for a loss.

On the more pristine reading of the original synchronic Dutch-book construction, the bookies in wooden shoes were only window dressing, and what was really being revealed (so the story went) was a structural inconsistency in the preferences of an agent who did not conform her degrees of belief to the probability calculus. In applying this reading to the diachronic setting, we need to divide cases. Consider first the case of an agent who eschews preset rules for changing degrees of belief. In this instance it is hard to see how the charge of inconsistency can legitimately be leveled. For how can such an agent's preferences over bets at t_1 be inconsistent with her preferences over bets at t_2 any more than her preferences over wines at t_1 can be inconsistent with her preferences over wines at t_2? Perhaps in response it will be urged that without melding together preferences at different times to form an integrated whole, it wouldn't be proper to speak of an enduring agent. That is certainly true, but surely the requirements for personal identity over time cannot be taken to entail rationality constraints—and conditionalization is allegedly such a constraint—since a person who behaves irrationally does not cease to be a person.

The agent who has adopted a rule for belief change is more open to the charge of inconsistency, since she has already committed herself at t_1 to what her preferences over bets will be at t_2. It would then seem that we can apply at t_1 the package principle introduced in the discussion of synchronic Dutch book: if an agent prefers each of a finite series of bets to the status quo, then she also prefers the package of bets to the status quo. To make this principle yield the desired consequence in the present setting, 'prefer' must be taken to mean prefer when the decision is viewed as an isolated one, which is the tacit understanding in effect when the critical odds for a bet on A are used to elicit the agent's degree of belief in A. But an agent who is not a conditionalizer can satisfy the package principle by taking 'prefer' to mean prefer when the decision to accept or reject the bet is placed in the context of a sequential decision problem. If we view the diachronic Dutch-book construction as a sequential decision process, the decision tree looks as in figure 2.1. The principles of rational decision making require that at decision node 1 the agent face up to what she knows about what her preferences will be at node 2, should she get there (see Seidenfeld 1988). She knows that at node 2 the tiniest premium will lead her to prefer to sell back to the bookie the bet on A, and she sees that in

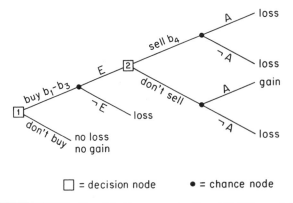

□ = decision node   ● = chance node

Figure 2.1
Diachronic Dutch book on a decision tree

the decision context this choice leads to a sure loss. She sees also that she gets to node 2 if at node 1 she chooses to buy b_1 to b_3 and E obtains, and further that if she chooses to buy b_1 to b_3 and E fails, she incurs a sure net loss. Thus, all things considered, she sees that buying b_1 to b_3 is unfavorable. It is on just these grounds that Maher (1992) maintains that the diachronic-Dutch-book argument is fallacious (see also Levi 1987).

To the extent that these decision-theoretic considerations are effective in undermining the diachronic-Dutch-book justification for conditionalization, they also bring into question the Dutch-book justification for the axioms of probability. In essence, the decision-theoretic message is to look before you leap. Such advice is just as valid in the synchronic setting as in the diachronic or multitemporal setting. And in the former setting, the advice clashes with the package principle needed in the argument for the principle of additivity of the probabilities of exclusive alternatives, which brings us full circle back to Schick's (1986) objection to Dutch-book arguments. The circle leaves me in an unsettled position. I agree, for example, that if I adopted a rule of belief change other than conditionalization and if I were cagey enough to draw up the decision tree for diachronic Dutch book, then I would refuse to accept the initial bets. But since I regard each of these bets as fair, should I not therefore recognize that there is something amiss in my opinion/preference structure? Grounds for a definitive answer do not exist, or if they do, I do not know of them.

A different and more modest justification for conditionalization has been given by Teller (1976), who argues that there are specifiable circumstances under which it can be maintained that if any change in belief is reasonable, then such a change must be via conditionalization. To identify some of these circumstances, Teller proves the following formal result (see also Teller and Fine 1975). Suppose that $Pr_{old}(E) > 0$ and that the agent's domain $\mathscr{A}$ of beliefs is *full* in the sense that for any number q and any $A \in \mathscr{A}$ such that $Pr_{old}(A) = r$ and $0 \leqslant q \leqslant r$ there is a $B \in \mathscr{A}$ such that $B \models A$ and $Pr_{old}(B) = q$. Suppose further that $Pr_{new}(\cdot)$ is such that $Pr_{new}(E) = 1$ and that for all $A, B \in \mathscr{A}$ such that $A \models E$ and $B \models E$, if $Pr_{old}(A) = Pr_{old}(B)$, then $Pr_{new}(A) = Pr_{new}(B)$. Then $Pr_{new}(\cdot) = Pr_{old}(\cdot/E)$.

As can easily be verified under the assumption that $Pr_{old}(E) = 0$, Teller's crucial condition $C(E)$ is equivalent to $C'(E)$:

$C(E)$ For all $A, B \in \mathscr{A}$ such that $A \models E$ and $B \models E$, if $Pr_{old}(A) = Pr_{old}(B)$, then $Pr_{new}(A) = Pr_{new}(B)$.

$C'(E)$ For all A, $B \in \mathcal{A}$ (whether or not they entail E), if $\mathrm{Pr_{old}}(A/E) = \mathrm{Pr_{old}}(B/E)$, then $\mathrm{Pr_{new}}(A) = \mathrm{Pr_{new}}(B)$.

There are clear cases where we want to impose $C(E)$ or $C'(E)$ for at least some A and B. Thus, let A be the proposition that Dancer will win the Derby, B the proposition that Prancer will win the Derby, and E the proposition that one or the other has won. Suppose that an agent is initially equally confident of A and B. She now learns precisely that E— that and no more. It would seem that, in accord with $C(E)$, she would be unreasonable in these circumstances to adjust her degrees of belief so that Dancer is now preferred to Prancer (or vice versa). But to invoke the formal result, we need to extend the argument to all pairs of initially equally probable propositions entailing E. It is hard to see how this can be done for any particular $\mathcal{A}$ that is sufficiently rich without using reasoning that would apply equally to any $\mathcal{A}$ and would thus abandon the modesty of the approach.

The basis for an immodest justification can perhaps be found in van Fraassen's (1989) result that under the assumption of the fullness of $\mathcal{A}$, $C(E)$ is implied by the requirement that the new probability of any proposition $A \in \mathcal{A}$ is a function solely of the evidence E and the old probability of A. It is well to note, however, that van Fraassen himself would not take such a justification to imply that conditionalization is necessary for rationality, since in his view rationality does not require that belief change follows a preset rule (see van Fraassen 1989 and 1990).

A different motivation for Jeffrey conditionalization starts from the idea that one should make as small a change as possible in one's overall system of beliefs compatible with the shift in those beliefs directly affected by the learning experience. Consider a probability function Pr on $\mathcal{A}$, thought of as giving the probabilities prior to making an observation. Let $\{E_i\}$ be a partition, intended as the locus of belief change, and let Pr* be a measure on $\{E_i\}$ such that $\mathrm{Pr}^*(E_i) > 0$ and $\sum_i \mathrm{Pr}^*(E_i) = 1$, intended to give the new probabilities of the E_i after observation. One would like to extend Pr* to a probability measure Pr** on $\mathcal{A}$ in such a way that Pr** makes as minimal a change as possible in Pr. Relative to several natural distance measures, the probability obtained by Jeffrey conditionalization fits the bill, although for some distance measures it may not do so uniquely (see Diaconis and Zabell 1982).

When the effect of observation is not so simple as to be localizable in a single partition, the method for updating probabilities becomes problematic. Suppose that one's experience results in new degrees of belief for each of the partitions $\{E_i\}$ and $\{F_j\}$. It is not guaranteed a priori that these degrees of belief are mutually coherent in the sense that they are extendible to a full probability on $\mathscr{A}$. A necessary and sufficient condition for the existence of such an extension is supplied by Diaconis and Zabell (1982). Assuming coherence, one could proceed to produce a new probability function by successive Jeffrey conditionalizations on the two partitions. But the order of conditioning may matter. If we denote the results of Jeffrey conditionalizing on $\{E_i\}$ (respectively $\{F_j\}$) by $\mathrm{Pr}_E(\cdot)$ ($\mathrm{Pr}_F(\cdot)$), then the order does not matter in that $\mathrm{Pr}_{EF}(\cdot) = \mathrm{Pr}_{FE}(\cdot)$ just in case $\mathrm{Pr}_F(E_i) = \mathrm{Pr}(E_i)$ and $\mathrm{Pr}_E(F_j) = \mathrm{Pr}(F_j)$ for all i and j.[12] The interested reader is referred to Diaconis and Zabell 1982 and van Fraassen 1989 for more discussion of these and related matters.

While the cumulative weight of the various justifications for conditionalization seems impressive, it should be noted that the starting assumptions of strict and Jeffrey conditionalization are left untouched. The former assumes that learning experiences have a precise propositional content in the sense that there is a proposition E that captures everything learned in the experience, while the latter assumes that if there is no precise propositional content, still the resulting belief changes can be localized to a partition. One or the other of these assumptions is surely correct for an interesting range of cases, but it is doubtful that they apply across the board. And where the doubt is realized, the present form of Bayesianism is silent.

In the remainder of this book I will concentrate on cases where strict conditionalization applies.

7   Lewis's Principal Principle

What David Lewis (1980, 1986) calls the principal principle (PP) may be viewed both as a rationality constraint on personal probabilities and as an implicit definition of objective probabilities. To paraphrase Lewis, (PP) requires that if $\mathrm{Pr}(\cdot)$ is a rational degree of belief function, A a proposition asserting that some specified event occurs at time t (e.g., a given coin lands heads up when flipped at t), A_p the proposition that asserts that the chance or objective probability at time t of A's holding is p, and E any proposition compatible with A that is admissible at t, then $\mathrm{Pr}(A/A_p \ \& \ E) = p$. Admissibility is, as Lewis notes, a tricky notion. But for present purposes it suffices to focus on one category of evidence that should be admissible in the intended sense, namely, any proposition E about matters of particular historical fact up to time t (e.g., information about the outcomes of past flips of the coin).

A glance at Bayes's calculations reported in chapter 1 is enough to establish that the Reverend Thomas himself used a version of (PP). Some of the mathematical niceties of Bayes's application of (PP) will be taken up in chapter 4, but these will be ignored in the present chapter to simplify the discussion.

Some early critics of probabilistic epistemology worried that the standard probability apparatus doesn't suffice to capture the full force of uncertain judgments. Consider two cases of partial knowledge. In the first I know literally nothing about a coin, save that it is two-sided and has a head and a tail. In the second I learn that 10,000 flips have produced 5,023 heads. If A is the proposition that the next flip will be heads, then in each of the two cases my degree of belief conditional on the total available evidence will presumably be (roughly) .5. But in the second case the "weight" of the evidence seems much greater, and consequently, my degree of belief is much firmer. The worry is that two numbers are needed to characterize my belief state, one describing my degree of belief, the other describing the weight of the evidence. But by using (PP), we can show that information about weight is already encoded in the standard probabilities. If we assume for sake of convenience that p can take on only discrete values p_i, we can write

$$\Pr(A/E) = \sum_i \Pr(A/A_{p_i} \,\&\, E) \times \Pr(A_{p_i}/E)$$

$$= \sum_i p_i \times \Pr(A_{p_i}/E),$$

where the first equality uses the principle of total probability and the second follows by (PP). The probability of A on E is thus the first moment of the distribution $\Pr(A_{p_i}/E)$. One would expect this distribution to look like figure 2.2a in the first hypothesized case and like figure 2.2b in the second. Thus, as E. T. Jaynes (1959) suggests, at least part of what is meant by 'weight of evidence' can be explicated in terms of the concentration of the $\Pr(A_p/E)$ distribution. This sense of weight is connected to the notion of firmness or resiliency, since presumably the greater the weight, the more new information about the outcomes of additional coin flips that is needed to significantly alter $\Pr(A_{p_i}/E)$ and thus $\Pr(A/E)$.

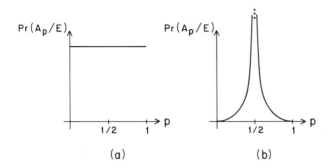

Figure 2.2
The distribution of personal probability of the objective probability

Principle (PP) is also (jokingly) referred to as Miller's principle because David Miller (1966) claimed to show that the principle is inconsistent. There is no need to review Miller's attack here, since Jeffrey (1970) and Howson and Urbach (1989) have successfully parried the attack. But it is worth reviewing van Fraassen's assessment of (PP), since if he is correct, (PP) has a less lofty status than it might seem to have at first glance.

Van Fraassen writes,

The intuition that Miller's Principle is a requirement of rationality firmly links its credentials to a certain view of ourselves—namely, that we are finite, temporally conditioned rational beings. We have no crystal balls, and no way to gather information about the future which goes beyond the facts which have become settled to date. If we thought instead that Miller's Principle must apply to all possible and conceivable rational beings, we would have to conclude that omniscience implies determinism. (1989, p. 196)

The argument proceeds by supposing that there is a rational agent who is omniscient. For that agent, $Pr(A) = 1$ or 0, according as A is true or false for any proposition A. So $Pr(A_{p*}) = 1$ for some unique $0 \leqslant p^* \leqslant 1$. But then by (PP), we get $Pr(A/A_{p*}) = p^* = Pr(A \,\&\, A_{p*})/Pr(A_{p*}) = Pr(A)$. Hence p^* is 1 or 0, according as A is true or false, which is determinism.

There is pressure, however, to argue the other way around. The objective chance of a specified outcome (e.g., the reflection of a photon by a half-silvered mirror) is $0 < p^* < 1$. This I know because quantum mechanics (QM) tells me so. Therefore, if I am rational, I shouldn't assign 1 or 0 as my degree of belief in the outcome. So objective chance is incompatible with rational omniscience about the future, and the scope of (PP) does after all include all rational agents. A potential difficulty with this line is that it might seem to conflict with a reliability conception of knowledge. Thus, suppose that a person is able to correctly predict the future time after time. Wouldn't we eventually be willing to say that this person knows what the future holds? Correct prediction in itself is not a sure indicator of the relevant sort of reliability, for it is consistent with lucky guessing. What is needed for knowledge is the existence of a belief-forming mechanism that reliably yields certainty, a probability of 1 or 0, on the events in question. But such a mechanism is arguably inconsistent with the seemingly irreducible nontrivial probabilities involved in quantum events. Indeed, several of the no-go results for hidden-variable interpretations of QM are not so much proofs that no deterministic mechanism underlies QM as they are demonstrations of the inconsistency of treating quantum-mechanical magnitudes as if they had simultaneously determinate values.

In other cases, such as classical statistical mechanics, we want to maintain both determinateness and determinism on the microscopic level and yet speak of objective chances of events defined on the macroscopic level. For example, if a gas is initially confined to one half of a container by a partition and the partition is removed, then the chance is overwhelmingly great that in a time short by macroscopic standards the gas molecules, insofar as macroscopic measurements will be able to ascertain, will become evenly distributed over the entire container. In assigning personal probabilities, it would be irrational to ignore such teachings of statistical mechanics. But this judgment, as opposed to the parallel judgment in the QM case, rests, as van Fraassen says, on our view of ourselves as temporally bounded agents who have no crystal balls for reading the future. And it also rests on our view of ourselves as being bounded in other ways as well, in particular, as being unable to discern the current exact microstate of the gas. This second limitation means, in effect, that the admissible evidence for (PP) is more circumscribed than originally announced, which is an indication that the probabilities provided by classical statistical mechanics are not wholly objective. This is not an unwelcome conclusion, since it is generally acknowledged that these probabilities are partly physical and partly epistemic.

Principle (PP) also has the apparent virtue that when combined with the law of large numbers, it explains how we can come to learn the values of objective chance parameters. Consider a coin-flipping case with independent and identically distributed (IID) trials and an objective chance p for heads. Starting from p, we can construct (as explained in appendix 2) a measure $\mathscr{P}\iota$ on subsets of the collection of all possible outcomes of an infinite repetition of this chance experiment, and we can prove that the $\mathscr{P}\iota$ measure of the set of all infinite sequences of flips in which the relative frequency of heads converges to p is 1 (the strong law of large numbers). It follows that if one starts by assigning a nonzero prior to the hypothesis that the objective chance of heads is p, obeys (PP), and updates probabilities by conditionalizing on the outcomes of repeated coin flips, then in almost every infinite repetition of the experiment (i.e., except for a set of $\mathscr{P}\iota$-measure 0) one's personal probability will converge to 1 on the said hypothesis in the limit as the number of flips goes to infinity (see chapter 4).

The mathematics here is impeccable, but the metaphysics remains murky. If we think of p as something like a single-case propensity, then the original application of (PP) has a plausible ring to it. Moreover, given the assumption of IID trials, the objective probability in n trials of getting m heads is

$$\binom{n}{m} p^m (1 - p)^{n-m}.$$

It follows that as $n \to \infty$ the objective probability goes to 0 that the relative frequency of heads differs from p by any specified $\varepsilon > 0$ (a form of the weak law of large numbers). So by applying (PP) at each stage of this reasoning, we can conclude that our personal probability goes to certainty that the frequency of heads comes within any desired $\varepsilon > 0$ of the true objective probability. But to get a personal-probability analogue of the strong form of the large numbers, we need to operate with the measure $\mathscr{P}\iota$ on the collection of infinite repetitions, and it is not immediately apparent why $\mathscr{P}\iota$ should function as an objective probability in the relevant sense of (PP), that is, so as to underwrite the conclusion that our personal probability ought to be one that in this infinite repetition of the experiment the limit of the relative frequency of heads will equal the objective probability of heads. The original (PP) can be defended on the grounds that it is constitutive of what is meant by objective probability. But we can only get away with such a move once.

Rather than start with single-case probabilities and then build the measure $\mathscr{P}\iota$ on sets of infinite sequences, J. L. Doob (1941) proposed that we take as basic the measure $\mathscr{P}\iota$, identify the event (say) of a coin's landing heads on the 35th flip with the set of all infinite sequences that yield heads in the 35th place, and then take the $\mathscr{P}\imath$ measure of this collection to be the probability of heads on the 35th flip (and thus by the IID assumption, the probability of heads on any trial). But what is now lacking is the conviction that this probability value functions enough like a single-case propensity so as to underwrite (PP) as applied to a particular, concrete flip.

It remains to be seen whether the difficulty here is trying to tell us something about the strong law of large numbers or about (PP) or both.

8 Descriptive versus Normative Interpretations of Bayesianism

Is Bayesianism to be regarded as descriptive of actual reasoning, or does it rather fix the pathways that "correct" or "rational" inductive reasoning must follow? Bayes's arguments for the probability axioms and the modern descendants of these arguments, the Dutch-book construction, certainly presuppose a normative aim, as do the discussions of rules of conditionalization and Lewis's (PP). This is just as well, since it is currently a matter of lively controversy as to whether actual agents can be represented as obeying the Bayesian constraints on belief and the allied decision rule of maximizing expected utility (see Kahneman, Slovic, and Tversky 1982).

Eschewing the descriptive in favor of the normative does not erase all difficulties. For 'ought' is commonly taken to imply 'can', but actual inductive agents can't, since they lack the logical and computational powers required to meet the Bayesian norms. The response that Bayesian norms should be regarded as goals toward which we should strive even if we always fall short is idle puffery unless it is specified how we can take steps to bring us closer to the goals. To make the complaint concrete, note that in a rich language, agents who are computationally bounded may fail to satisfy probability axiom (A2). This failure is not a mere inadvertence that can easily be corrected, since by their very nature these agents fall short of the logical omniscience that requires recognition of all logical truths in the domain of Pr. Thus a realistic Bayesianism must somehow make room for logical learning. And it is in this regard that one must agree with Good (1977) that probability qua degree of belief can change not only as the result of observation and experiment but also as a result of calculation and pure thought. This matter will surface again in chapter 5 in the discussion of the problem of old evidence.

Actual agents also fall short of logical omniscience by being unable to parse all the possibilities, and this inability can skew degrees of belief.

The probability calculus requires that the degrees of belief assigned to Einstein's general theory of relativity (GTR) and its negation sum to one. But when Einstein first proposed GTR, physicists had only the dimmest idea of what was contained in the portion of possibility space denoted by ¬GTR, and thus their assessments of the probability of GTR were ill informed in the worst way. 'Explore the space of possibilites' is an empty injunction unless accompanied by practical guidelines. Although I have no general prescription to offer in this regard, I will offer in chapter 7 some examples of how the exploration has been conducted in some actual and challenging cases. It should be noted, however, that such an exploration cannot be undertaken in an orthodox Bayesian fashion, for the recognition of heretofore obscured possibilities is typically accompanied by belief changes, and it is hardly possible to account for all of these changes by conditionalization, whether of the strict or Jeffrey form. This matter will be taken up in chapters 7 and 8.

Finally, the considerations of chapter 9 raise a new and different challenge to the normative status of Bayesianism by showing that the structural constraints it imposes on degrees of belief entail a substantive knowledge of a kind that most scientists would not regard as appropriate to bring to a domain of inquiry.

9   Prior Probabilities

The topic of priors will come up again and again in the chapters below. While it would not be productive to anticipate in advance all of the nuances of the discussion, it may nevertheless be useful to outline the shape of one of the central issues. For the Bayesian apparatus to be relevant to scientific inference, it seems that what it needs to deliver are not mere subjective opinions but reasonable, rational, objective degrees of belief. Thence comes the challenge: How are prior probabilities to be assigned so as to make this delivery possible? (Note that the presupposition of this challenge is that the other factors involved in Bayes's theorem, the likelihoods, are unproblematic in a way that priors are not. While this may be true in some special cases, it is most certainly not true in general. But since this point only serves to complicate the matter at hand, I waive it for the time being.)

Three responses to the challenge are to be found in the Bayesian corpus. The first is that the assignment of priors is not a critical matter, because as the evidence accumulates, the differences in priors "wash out." Chapter 6 will examine in detail various theorems that are supposed to demonstrate this washout effect. In advance it is fair to say that the formal results apply only to the long run and leave unanswered the challenge as it applies to the short and medium runs.

The second response is to provide rules to fix the supposedly reasonable initial degrees of belief. In chapter 1 we met Thomas Bayes's attempt to justify the rule of a uniform prior distribution. We saw that, although ingenious, Bayes's attempt is problematic. Other rules for fixing priors suffer from similar difficulties. And generally, none of the rules cooked up so far are capable of coping with the wealth of information that typically bears on the assignment of priors.

The third response is that while it may be hopeless to state and justify precise rules for assigning numerically exact priors, still there are plausibility considerations that can be used to guide the assignments. By way of concrete illustration, consider the recent controversy about AIDS transmission raised by Lorraine Day, a San Francisco surgeon. Day worried that surgeons might contract AIDS by inhaling the air-borne blood of infected patients. To protect against this risk, she urged her colleagues to wear space-suit-like outfits when using high speed drills and saws that create a fine mist of blood droplets. Critics responded that it is implausible that Day's hypothetical transmission mechanism poses any serious risk, since the AIDS virus should remain suspended in the blood droplets and since these droplets are typically too large to pass through the openings of standard surgical masks. In Bayesian jargon, the critics are urging that these plausibility considerations justify assigning a low prior to Day's hypothesis.

This third response does point to an important aspect of actual scientific reasoning, but at the same time it opens the Bayesians to a new challenge, which Fredrick Suppe has put in the form of a dilemma:

If standard inductive logic [i.e., Bayesianism] is intended to provide an analysis of that plausibility reasoning, then we have a vicious regress where each iteration of the Bayesian method requires a logically prior application; hence it is impossible to ever get the Bayesian method going. Hence standard inductive logic is an inadequate model of scientific reasoning about evidence and the evaluation of hypotheses. If, on the other hand, standard inductive logic does not provide an analysis of that plausibility reasoning, standard inductive logic is a critically incomplete, hence an inadequate model of scientific reasoning about evidence and the evaluation of hypotheses. (1989, p. 399)

Although some Bayesians have tried to seize one or the other of the horns of this dilemma, it seems to me that the only escape is between the horns.[13] That is, Bayesians must hold that the appeal to plausibility arguments does not commit them to the existence of a logically prior sort of reasoning: plausibility assessment. Plausibility arguments serve to marshall the relevant considerations in a perspicuous form, yet the assessment of these considerations comes with the assignment of priors. But, of course, this escape succeeds only by reactivating the original challenge. The upshot seems to be that some form of the washout solution had better work not just for the long run but also for the short and medium runs as well.

The matter of plausibility arguments also serves to bring to the surface one of the lingering doubts that many philosophers have about Bayesianism. The worry is that the Bayesian apparatus is just a kind of tally device used to represent a more fundamental sort of reasoning whose essence does not lie in the assignment of little numbers to propositions in accord with the probability axioms. The only effective way to assuage this worry is to examine the many attempts to capture scientific reasoning in non-Bayesian terms and to detail how each of these attempts fails. Some of this work will be done in chapter 3.

10 Conclusion

In the next two chapters I will assume that Bayesians are armed with the probability calculus, including countable additivity if it should prove helpful, and also with whatever form of conditionalization seems appropriate to the context. How this arsenal is deployed to attack problems in confirmation theory will be the subject of discussion.

Appendix 1: Conditional Probability

Conditional probability

Suppose that $\mathscr{B} \subseteq \mathscr{A}$ and let $\mathscr{B}^0$ stand for the noncontradictory elements of $\mathscr{B}$. Then a conditional probability $\Pr(\cdot/\cdot)$ is a function from $\mathscr{A} \times \mathscr{B}^0$ to $\mathbb{R}$ satisfying the following:

CP1 $\Pr(\cdot/B)$ is an unconditional probability on $\mathscr{A}$ for any $B \in \mathscr{B}^0$.

CP2 $\Pr(B/B) = 1$ for any $B \in \mathscr{B}^0$.

CP3 $\Pr(A \,\&\, B/C) = \Pr(B/C) \times \Pr(A/B \,\&\, C)$ for any $A \in \mathscr{A}$, $B \in \mathscr{B}^0$, $C \in \mathscr{B}^0$, and $B \,\&\, C \in \mathscr{B}^0$.

$\Pr(\cdot/\cdot)$ is said to be *full* just in case $\mathscr{B} = \mathscr{A}$. It will be assumed here that we are dealing with full conditional probabilities, since any conditional probability can be extended to a full one. If $\models N$ for $N \in \mathscr{B}^0$, $\Pr(\cdot) \equiv \Pr(\cdot/N)$ is the unconditional probability associated with $\Pr(\cdot/\cdot)$. ($\Pr(\cdot)$ is independent of the choice of N.) It is easy to see that if $B \in \mathscr{B}^0$ is such that $\Pr(B) \neq 0$, $\Pr(A/B) = \Pr(A \,\&\, B)/\Pr(B)$ for any $A \in \mathscr{A}$.

Countable additivity, disintegrability, and conglomerability

A *partition* of the possibilities consists of a set $\{H_1, H_2, \ldots\}$ of statements $H_i \in \mathscr{A}$ that are pairwise exclusive and mutually exhaustive, i.e., $\{H_i, H_j\} \models P \,\&\, \neg P$ for $i \neq j$ and $\{\neg H_1, \neg H_2, \ldots\} \models P \,\&\, \neg P$. Let $\Pr(\cdot/\cdot)$ be a conditional probability and $\Pr(\cdot)$ its associated unconditional probability. Here $\Pr(\cdot)$ is said to be *countably additive* just in case the continuity condition (C) (p. 36) holds. This condition implies that for any partition $\{H_1, H_2, \ldots\}$, $\lim_{n \to \infty} \Pr(\bigvee_{i \leqslant n} H_i) = \sum_{i=1}^{\infty} \Pr(H_i) = 1$.

A countably additive $\Pr(\cdot)$ associated with the conditional probability $\Pr(\cdot/\cdot)$ has the property of *disintegrability*: for any $A \in \mathscr{A}$ and any partition $\{H_1, H_2, \ldots\}$, $\Pr(A) = \sum_{i=1}^{\infty} \Pr(A/H_i) \times \Pr(H_i)$. Disintegrability for a partition in turn entails *conglomerability*: if $k_1 \leqslant \Pr(A/H_i) \leqslant k_2$ for every i, then $k_1 \leqslant \Pr(A) \leqslant k_2$. The circle closes: conglomerability with respect to every countable partition implies countable additivity (Schervish, Seidenfeld, and Kadane 1984).

The failure of countable additivity and consequently of conglomerability can lead to very awkward situations, such as the failure of a natural principle of dominance, which demands that if an action O_1 is conditionally preferred to O_2 for each member of a partition $\{H_1, H_2, \ldots\}$, then O_1 is unconditionally preferred to O_2 (see Kadane, Schervish, and Seidenfeld 1986). However, nonconglomerability should not be allowed to become a bugaboo, since even when countable additivity holds, conglomerability can fail for uncountable partitions (see Kadane, Schervish, and Seidenfeld 1986).

The failure of disintegrability is also very awkward for Bayesian inference problems, since it means that the denominator of Bayes's theorem (2.1) cannot be written in the form given in (2.2); so, for example, the probability of the experimental outcome E cannot be assessed in terms of how well the alternative hypotheses H_i explain the outcome (as given by the likelihoods $\Pr(E/H_i \ \& \ K)$) and how antecedently probable the hypotheses are (as given by the priors $\Pr(H_i/K)$). This theoretical worry is of no practical importance if in all realistic cases inference involves only a finite number of H_i.

Finally, it should be noted that Thomas Bayes's own calculations (chapter 1) implicitly assumed a form of countable additivity (see chapter 4).

Appendix 2: Laws of Large Numbers

Recall that a finite field $\mathscr{F}_0$ over the set Ω is a collection of subsets of Ω that contains Ω and is closed under complementation and finite unions (and thus under finite intersections). A σ field $\mathscr{F}_\sigma$ is closed under countable unions (and thus under countable intersections). Each $\mathscr{F}_0$ generates a σ field $\mathscr{F}_\sigma$, namely, the smallest σ field containing $\mathscr{F}_0$. Let $p\!\imath$ be a finitely additive probability measure on $\mathscr{F}_0$. (For any $A \in \mathscr{F}_0, p\!\imath(A) \geqslant 0$; $p\!\imath(\Omega) = 1$. And for any $A_1, A_2, \ldots, A_n \in \mathscr{A}$ such that $A_i \cap A_j = \varnothing$ for $i \neq j$, $p\!\imath(\bigcup_{i=1}^{n} A_i) = \sum_{i=1}^{n} p\!\imath(A_i)$.) The function $p\!\imath$ is said to be continuous from above at $\varnothing$ just in case if $A_i \in \mathscr{A}$, $i = 1, 2, \ldots$, are such that $A_{i+1} \subseteq A_i$ and $\bigcap_{i=1}^{\infty} A_i = \varnothing$, then $p\!\imath(\bigcup_{i=1}^{n} A_i) \to 0$ as $n \to \infty$. (This implies a conditional form of countable additivity: if $\bigcup_{i=1}^{\infty} B_i \in \mathscr{A}$ and the B_i are pairwise disjoint, then $p\!\imath(\bigcup_{i=1}^{\infty} B_i) = \sum_{i=1}^{\infty} p\!\imath(B_i)$.) Carathéodory's extension lemma shows that for such a $p\!\imath$ there is a unique extension to a countably additive probability measure $\mathscr{P}\!\imath$ on the σ field generated by $\mathscr{F}_0$.

In the application to IID trials of coin flips, take Ω to be the collection of all one-sided infinite sequences of possible outcomes. (So a typical $\omega \in \Omega$ would be HTHHTHTTT....) Define a finite field of subsets of Ω by starting with the "cylinder sets," where a cylinder set is the set of all ω's that agree on the outcomes in a finite number of places. (A typical cylinder set would be the collection of all ω's that have heads in the 20th place and tails in the 801st place.) $\mathscr{F}_0$ is then the finite field consisting of the empty set and finite disjoint unions of the cylinder sets. A $p\!\imath$ measure is defined on $\mathscr{F}_0$ by assigning probabilities to the cylinder sets in the natural way. For example, the measure of the set of all ω's having heads in the 32nd and 41st places and tails in the 33rd, 58th, and 105th places is $p^2(1 - p)^3$, where p is the objective probability of heads. Since this $p\!\imath$ is continuous from above at $\varnothing$, it follows from Carathéodory's lemma that there is a unique countably additive extension $\mathscr{P}\!\imath$ of $p\!\imath$ to the σ field $\mathscr{F}_0$ generated by $\mathscr{F}_0$.

If we now let $j_n(\omega)$ stand for the number of heads in the first n trials of ω, the weak and strong form of the law of large numbers can be stated as follows:

WLLN   The $\mathscr{P}\imath$ measure of the set of ω's for which $|(j_n(\omega)/n) - p| > \varepsilon$ approaches 0 as $n \to \infty$ for any $\varepsilon > 0$.

SLLN   The $\mathscr{P}\imath$ measure of the set of all ω's such that $\lim_{n \to \infty} (j_n(\omega)/n) \neq p$ is 0.

To put (SLLN) in its positive form, the Pr probability is one that the limiting relative frequency of heads converges to p.

As indicated in section 7, a form of the weak law of large numbers can be formulated and proved without the help of countable additivity. Roughly, for any $\varepsilon > 0$, the probability (in the objective sense or in the degree-of-belief sense tempered by Lewis's principal principle) that the actually observed relative frequency of heads differs from p by more than ε goes to 0 as the number of flips goes to infinity. This form of the law of large numbers is to be found in the work of Bernoulli. The strong form of the law of large numbers, which requires countable additivity, was not proved until this century (see Billingsley 1979 for a proof).

3 Success Stories

The successes of the Bayesian approach to confirmation fall into two categories. First, there are the successes of Bayesianism in illuminating the virtues and pitfalls of various approaches to confirmation theory by providing a Bayesian rationale for what are regarded as sound methodological procedures and by revealing the infirmities of what are acknowledged as unsound procedures. The present chapter reviews some of these explanatory successes. Second, there are the successes in meeting a number of objections that have been hurled against Bayesianism. The following chapter discusses several of these successful defenses. Taken together, the combined success stories help to explain why many Bayesians display the confident complacency of true believers. Chapters 5 to 9 will challenge this complacency. But before turning to the challenges, let us give Bayesianism its due.

1 Qualitative Confirmation: The Hypotheticodeductive Method

When Carl Hempel published his seminal "Studies in the Logic of Confirmation" (1945), he saw his essay as a contribution to the logical empiricists' program of creating an inductive logic that would parallel and complement deductive logic. The program, he thought, was best carried out in three stages: the first stage would provide an explication of the qualitative concept of confirmation (as in 'E confirms H'); the second stage would tackle the comparative concept (as in 'E confirms H more than E' confirms H'''); and the final stage would concern the quantitative concept (as in 'E confirms H to degree r'). In hindsight it seems clear (at least to Bayesians) that it is best to proceed the other way around: start with the quantitative concept and use it to analyze the comparative and qualitative notions. The difficulties inherent in Hempel's own account of qualitative confirmation will be studied in section 2. This section will be devoted to the more venerable hypotheticodeductive (HD) method.

The basic idea of HD methodology is deceptively simple. From the hypothesis H at issue and accepted background knowledge K, one deduces a consequence E that can be checked by observation or experiment. If Nature affirms that E is indeed the case, then H is said to be HD-confirmed, while if Nature affirms $\neg E$, H is said to be HD-disconfirmed. The critics of HD have so battered this account of theory testing that it would be unseemly to administer any further whipping to what is very nearly a dead horse.[1] Rather, I will review the results of the jolly Bayesian postmortem.

Suppose that (a) $\{H, K\} \models E$, (b) $0 < \Pr(H/K) < 1$, and (c) $0 < \Pr(E/K) < 1$.[2] Condition (a) is just the basic HD requirement for confirmation. Condition (b) says that on the basis of background knowledge K, H is not known to be almost surely true or to be almost surely false, and (c) says likewise for E. By Bayes's theorem and (a), it follows that

$$\Pr(H/E \ \& \ K) = \Pr(H/K)/\Pr(E/K). \tag{3.1}$$

By applying (b) and (c) to (3.1), we can conclude that $\Pr(H/E \ \& \ K) > \Pr(H/K)$, i.e., *E incrementally confirms H* relative to K. Thus Bayesianism is able to winnow a valid kernel of the HD method from its chaff.

(To digress, this alleged success story might be questioned on the grounds that HD testing typically satisfies not condition (a) but rather a condition Hempel calls the "prediction criterion" of confirmation; namely, (a') E is logically equivalent to $E_1 \ \& \ E_2$, $\{H, K, E_1\} \models E_2$, but $\{H, K\} \not\models E_2$. That is, HD condition (a) is satisfied with respect to the conditional prediction $E_1 \rightarrow E_2$, but the total evidence consists of E_1 and E_2 together. Let us use Bayes's theorem to draw out the consequences of (a'). It follows that $\Pr(H/E_1 \ \& \ E_2 \ \& \ K) = \Pr(H/E_1 \ \& \ K)/\Pr(E_2/E_1 \ \& \ K)$. Thus if $\Pr(E_2/E_1 \ \& \ K) < 1$ and $\Pr(H/E_1 \ \& \ K) = \Pr(H/K)$, the total evidence $E_1 \ \& \ E_2$ incrementally confirms H. These latter two conditions are satisfied in typical cases of HD testing. For example, let H be Newton's theory of planetary motion, let E_1 be the statement that a telescope is pointed in such and such a direction tomorrow at 3:00 P.M., and let E_2 be the statement that Mars will be seen through the telescope. Presumably, E_1 is probabilistically irrelevant to the theory, and E_2 is uncertain on the basis of E_1 and K.)

Notice also that from (3.1) it follows that the smaller the value of the prior likelihood $\Pr(E/K)$, the greater the incremental difference $\Pr(H/E \ \& \ K) - \Pr(E/K)$, which seems to validate the saying that the more surprising the evidence is, the more confirmational value it has. This observation, however, is double-edged, as we will see in chapter 5.

The problem of irrelevant conjunction, one of the main irritants of the HD method, is also illuminated. If $\{H, K\} \models E$, then also $\{H \ \& \ I, K\} \models E$, where I is anything you like, including a statement to which E is, intuitively speaking, irrelevant. But according to the HD account, E confirms $H \ \& \ I$. In a sense, the Bayesian analysis concurs, since if $\Pr(H \ \& \ I/K) > 0$, it follows from the reasoning above that E incrementally confirms $H \& I$. However, note that it follows from (3.1) that the amounts of incremental confirmation that H and $H \& I$ receive are proportional to their prior probabilities:

$$\Pr(H/E \& K) - \Pr(H/K) = \Pr(H/K)[(1/\Pr(E/K)) - 1]$$

$$\Pr(H \& I/E \& K) - \Pr(H \& I/K) = \Pr(H \& I/K)[(1/\Pr(E/K)) - 1].$$

Since in general $\Pr(H \& I/K) < \Pr(H/K)$, adding the irrelevant conjunct I to H lowers the incremental confirmation afforded by E.

Finally, it is worth considering in a bit more detail the case of HD disconfirmation. Thus, suppose that when Nature speaks, she pronounces $\neg E$. If $\{H, K\} \models E$ and if K is held to be knowledge, then H must be false, so HD disconfirmation would seem to be equivalent to falsification. But as Duhem and Quine have reminded us, the deduction of observationally decidable consequences from high-level scientific hypotheses often requires the help of one or more auxiliary assumptions A. It is not fair to ignore this problem by sweeping the A's under the rug of K, since the A's are often every bit as questionable as H itself. Thus from Nature's pronouncement of $\neg E$ all that can be concluded from deductive logic alone is that $\neg H \vee \neg A$. If HD methodology were all there is to inductive reasoning, then there would be no principled way to parcel out the blame for the false prediction, and we would be well on the way to Duhem and Quine holism (see section 4 below). In particular, H could be maintained come what may if the only constraints operating were those that followed from direct observation and deductive logic. But the fact that the majority of scientists sometimes regard the maintenance of a hypothesis as reasonable and sometimes not is a fact of actual scientific practice that cries out for explanation. The Bayesian attempt at an explanation will be examined in section 7 below.

2 Hempel's Instance Confirmation

Having rejected the HD or prediction criterion of confirmation, Hempel constructed his own analysis of qualitative confirmation on a very different basis. He started with a number of conditions that he felt that any adequate theory of confirmation should satisfy, among which are the following:

Consequence condition   If $E \models H$, then E confirms H.

Consistency condition   If E confirms H and also H', then $\not\models \neg(H \,\&\, H')$.

Special consequence condition   If E confirms H and $H \models H'$, then E confirms H'.

Hempel specifically rejected the converse consequence condition:

Converse consequence condition   If E confirms H and $H' \models H$, then E confirms H'.

For to add the last condition to the first three would lead to the disaster that any E confirms any H.[3] (Note that HD confirmation satisfies the converse consequence condition but violates both the consistency condition and the special consequence condition.)

Hempel's basic idea for finding a definition of qualitative confirmation satisfying his adequacy conditions was that a hypothesis is confirmed by its positive instances. This seemingly simple and straightforward notion turns out to be notoriously difficult to pin down.[4] Hempel's own explication utilized the notion of the *development* of a hypothesis for a finite set I of individuals. Intuitively, $dev_I(H)$ is what H asserts about a domain consisting of just the individuals in I. Formally, $dev_I(H)$ for a quantified H is arrived at by peeling off universal quantifiers in favor of conjunctions over I and existential quantifiers in favor of disjunctions over I. Thus, for example, if $I = \{a, b\}$ and H is $(\forall x)(\exists y)Lxy$ (e.g., "Everybody loves somebody"), $dev_I(H)$ is $(Laa \vee Lab) \,\&\, (Lbb \vee Lba)$. We are now in a position to state the main definitions that constitute Hempel's account.

Definition E *directly Hempel-confirms* H iff $E \models dev_I(H)$, where I is the class of individuals mentioned in E.

Definition E *Hempel-confirms* H iff there is a class C of sentences such that $C \models H$ and E directly confirms each member of C.[5]

Definition E *Hempel-disconfirms* H iff E Hempel-confirms $\neg H$.

The difficulties with Hempel's account can be grouped into three categories. The first concerns the pillars on which the account was built: Hempel's so-called adequacy conditions. Bayesians have at least two ways of defining qualitative confirmation, one of which we already encountered in section 1; namely, E *incrementally confirms* H relative to K iff $Pr(H/E \,\&\, K) >$

$\Pr(H/K)$. The second is an absolute rather than incremental notion; specifically, *E absolutely confirms H* relative to K iff $\Pr(H/E \ \& \ K) \geqslant k > .5$. (A third criterion sometimes used in the literature, e.g., Mackie 1963, says that *E* confirms *H* relative to *K* just in case $\Pr(E/H \ \& \ K) > \Pr(E/K)$. The reader can easily show that on the assumption that none of the probabilities involved is zero, this *likelihood criterion* is equivalent to the incremental criterion.) In both instances there appears to be a mismatch, since Hempel's account is concerned with a two-place relation '*E* confirms *H*' rather than with a three-place relation ('*E* confirms *H* relative to *K*'). The Bayesians can accommodate themselves to Hempel either by taking K to be empty or by supposing that K has been learned and then working with the new probability function $\Pr'(\cdot) = \Pr(\cdot/K)$ obtained by conditionalization. But since one of the morals the Bayesians want to draw is that background knowledge can make a crucial difference to confirmation, I will continue to make K an explicit factor in the confirmation equation.

The first difficulty for Hempel's account can now be stated as a dilemma. For any choice of K compatible with H, Hempel's adequacy conditions accord well with the absolute notion of Bayesian confirmation. For example, if $\Pr(H/E \ \& \ K) > .5$ and $H \models H'$, then $\Pr(H'/E \ \& \ K) > .5$, so the special consequence condition is satisfied. But absolute confirmation cannot be what Hempel had in mind, since he holds that the observation of a single black raven a confirms the hypothesis that all ravens are black, even though for typical K's, $\Pr((\forall x)(Rx \to Bx)/Ra \ \& \ Ba \ \& \ K) \ll .5$. On the other hand, while the incremental concept of confirmation allows that a single instance can confirm a general hypothesis, both the consistency condition and the special consequence condition fail for not atypical K's, as examples by Carnap (1950) and Salmon (1975) show.[6] Of course, there may be some third probabilistic condition of confirmation that allows Hempel's account to pass between the horns of this dilemma. But it is up to the defender of Hempel's instance confirmation to produce the *tertium quid*. And even to conduct the search for a probabilistic *tertium quid* is to fall into the hands of the Bayesians.

The second category of difficulties revolves around the question of whether Hempel's account is too narrow. One reason for thinking so is that, as Hempel himself notes, a hypothesis of the form

$$(\forall x)(\exists y)Rxy \ \& \ (\forall x)(\forall y)(\forall z)[(Rxy \ \& \ Ryz) \to Rxz] \ \& \ (\forall x) \neg Rxx$$

cannot be Hempel-confirmed by any consistent E, since the development of such a hypothesis for a finite domain is inconsistent. Nor is the hypothesis $(\forall x)(\forall y)Rxy$ Hempel-confirmed by the set of evidence statements $\{Ra_ia_j\}$, where $i = 1, 2, \ldots, 10^9$ and $j = 1, 2, \ldots, 10^9 - 1$. Even more troublesome is the fact that Hempel's account is silent about how theoretical hypotheses are confirmed, for if, as Hempel intended, E is stated purely in the observational vocabulary and if H is stated in a theoretical vocabulary disjoint from the observational vocabulary, then E cannot, except in very uninteresting cases, Hempel-confirm H.[7] This silence is a high price to pay for overcoming some of the defects of the more vocal HD method.

Clark Glymour (1980) has sought to preserve Hempel's idea that hypotheses are confirmed by deducing positive instances of them from observation reports. In the case where H is stated in theoretical vocabulary, Glymour's bootstrapping method allows the deduction to proceed via auxiliary hypotheses, typically drawn from a theory T of which H itself is a part.[8] His basic confirmation relation is thus three-place: E confirms H relative to T.

The Bayesian response to these difficulties and to Glymour's reaction to them is twofold. First, there is no insuperable problem about how observational data can confirm, in either the incremental or absolute sense, a theoretical hypothesis; indeed, the application of Bayes's theorem shows just how such confirmation takes place, at least on the assumption that the prior probability of the hypothesis is nonzero (a matter that will be taken up in chapter 4). Second, unless bootstrap confirmation connects to reasons for believing the hypothesis or theory, it is of no interest. But once the connection is made, the bootstraps can be ignored in favor of the standard Bayesian account of reasons to believe. This matter will be examined in more detail in section 4 below.

The third category of difficulties is orthogonal to the second. Now the worry is that while Hempel's instance confirmation may be too narrow in some respects, it may be too liberal in other respects. Consider again the ravens hypothesis: $(\forall x)(Rx \rightarrow Bx)$. Which of the following evidence statements Hempel-confirm it?

E_1: Ra_1 & Ba_1

E_2: $\neg Ra_2$ & $\neg Ba_2$

E_3: $\neg Ra_3$

E_4: Ba_4

E_5: $\neg Ra_5$ & Ba_5

E_6: Ra_6 & $\neg Ba_6$

Only E_6 fails to Hempel-confirm the hypothesis, and that is because E_6 falsifies it. The indoor ornithology involved in using E_2 to E_5 as confirmation of the ravens hypothesis has struck many commentators as too easy to be correct. Bayesian treatments of Hempel's ravens paradox will be taken up in the following section.

If anything is safe in this area, it would seem to be that E_1 does confirm $(\forall x)(Rx \rightarrow Bx)$. But safe is not sure. Recall that Hempel's definition of confirmation is purely syntactical in that it is neutral to the intended interpretation of the predicates. This means that E_1 Hempel-confirms $(\forall x)(Rx \rightarrow Bx)$ even if we take Bx to mean not that x is black but that x is blite, i.e., x is first examined before the year 2000 and is black, or else is not examined before 2000 and is white. Let a_i be first examined in the year i. Then by the special consequence condition, Ra_1 & Ba_1 & Ra_2 & Ba_2 & ... & Ra_{1999} & Ba_{1999} Hempel-confirms the prediction $Ra_{2001} \rightarrow Ba_{2001}$, i.e., the prediction that if a_{2001} is a raven, then it is white, which is, to say the least, counterintuitive. We have here an instance of what Goodman (1983) calls the "new riddle of induction." The Bayesian treatment of this problem will be given in detail in chapter 4. But for now I will simply note on behalf of the Bayesians that they are not committed to assigning probabilities purely on the basis of the syntax of the hypothesis and the evidence, as Hempel's analogy between deductive and inductive logic would suggest. The present example is enough to show that an adequate account of confirmation must be sensitive to semantics, and this lesson is easily incorporated into Bayesianism.

3   The Ravens Paradox

In sections 1 and 2 Bayesianism gained reflected glory of sorts from the whippings the HD and Hempel accounts took. It is time for Bayesianism to earn additional glory of a more positive sort.

Hempel took it as a desirable consequence of his account that the evidence Ra & Ba confirms the hypothesis $(\forall x)(Rx \rightarrow Bx)$.[9] The paradox of the ravens in one of its forms arises from the fact that on Hempel's analysis, the evidence $\neg Rb$ & $\neg Bb$ also confirms $(\forall x)(Rx \rightarrow Bx)$. Before turning to the Bayesian analysis of the paradox itself, it is worth noting that the Bayesian is not even willing to go the first step with Hempel without first looking both ways.

Suppose that $0 < \Pr(H/K) < 1$, where H stands for the ravens hypothesis. Then by an application of Bayes's theorem it follows that finding a to be a black raven induces incremental confirmation,

$$\Pr(H/Ra \,\&\, Ba \,\&\, K) > \Pr(H/K),$$

just in case

$$\Pr(Ra/H \,\&\, K) > \Pr(Ra/\neg H \,\&\, K) \times \Pr(Ba/Ra \,\&\, \neg H \,\&\, K).$$

Incremental disconfirmation results just in case the inequality is reversed.[10] The reader is invited to reflect on the kinds of background knowledge K that will make or break these inequalities. Consider, for instance, a version of I. J. Good's (1967) example. We are supposed to know in advance (K) that we belong to one of two bird universes: one where there are 100 black ravens, no nonblack ravens, and 1 million other birds, or else one where there are 1,000 black ravens, 1 white raven, and 1 million other birds. Bird a is selected at random from all the birds and found to be a black raven. This evidence, Good claims, undermines the ravens hypothesis. Use the above formula to test this claim. Such exercises help to drive home the point that a two-place confirmation relation that ignores background evidence is not very useful.

Let us turn now to the Bayesian treatment of the bearing of the evidence of nonblack nonravens on the ravens hypothesis. Suppes (1966) invites us to consider an object a drawn at random from the universe. Set

$$\Pr(Ra \,\&\, Ba/K) = p_1, \qquad \Pr(Ra \,\&\, \neg Ba/K) = p_2,$$
$$\Pr(\neg Ra \,\&\, Ba/K) = p_3, \qquad \Pr(\neg Ra \,\&\, \neg Ba/K) = p_4. \tag{3.2}$$

Then

$$\Pr(\neg Ba/Ra \,\&\, K) = p_2/(p_1 + p_2) \tag{3.3}$$

and

$$\Pr(Ra/\neg Ba \,\&\, K) = p_2/(p_2 + p_4). \tag{3.4}$$

From (3.3) and (3.4) it follows that $\Pr(\neg Ba/Ra \,\&\, K) > \Pr(Ra/\neg Ba \,\&\, K)$ iff $p_4 > p_1$. But from what we know of the makeup of our universe, it seems safe to assume that $p_4 \gg p_1$, with the consequence that the conditional probability of a's being nonblack, given that it is a raven, is much greater than the conditional probability of a's being a raven, given that it is nonblack. The moral Suppes wants us to draw from this is that sampling from the class of ravens is more productive than sampling from the class of nonblack objects, since the former procedure is more likely to produce a counterexample to the ravens hypothesis.

There are two qualms about this moral. The first is that it doesn't seem directly useful to Bayesians; indeed, at first blush it seems more congenial to a Popperian line that emphasizes the virtues of attempted falsifications of hypotheses. Second, it is not clear how the moral follows from the inequality derived, since a was supposed to result from a random sample of the universe at large rather than from a random sample of either the class of ravens or the class of nonblack objects.

Horwich's (1982) attack on the ravens paradox starts from the observation that there are several ways to obtain the evidence $Ra \ \& \ Ba$, namely, to pick an object at random from the universe at large and find that it has both ravenhood and blackness, to pick an object at random from the class of ravens and find that it is black, or to pick an object at random from the class of black things and find that it is a raven. A similar remark applies to the evidence $\neg Rb \ \& \ \neg Bb$. Horwich introduces the notation R^*a to mean that a was drawn at random from the class of ravens and the notation $\neg B^*b$ to mean that b was drawn at random from the class of nonblack things. To illuminate the ravens paradox, he wants to compare the confirmational effects of the two pieces of evidence $R^*a \ \& \ Ba$ and $\neg B^*b \ \& \ \neg Rb$. According to Horwich's application of Bayes's theorem,

$$Pr(H/R^*a \ \& \ Ba \ \& \ K) = Pr(H/K)/Pr(R^*a \ \& \ Ba/K) \tag{3.5}$$

and

$$Pr(H/\neg B^*b \ \& \ \neg Rb \ \& \ K) = Pr(H/K)/Pr(\neg B^*b \ \& \ \neg Rb/K), \tag{3.6}$$

where K is the same as before. Thus

$$Pr(H/R^*a \ \& \ Ba \ \& \ K) > Pr(H/\neg B^*b \ \& \ \neg Rb \ \& \ K)$$

iff $\quad Pr(\neg B^*b \ \& \ \neg Rb/K) > Pr(R^*a \ \& \ Ba/K)$.

But the latter is true for our universe, Horwich asserts.

But as with Suppes's construction, it is not clear how this conclusion follows. In the first place, why is it true (as (3.5) and (3.6) assume) that

$$\Pr(R^*a \ \& \ Ba/H \ \& \ K) = \Pr(\neg B^*b \ \& \ \neg Rb/H \ \& \ K) = 1?$$

It is true that the probability of a randomly chosen raven being black, given $H \ \& \ K$, is 1, but $\Pr(R^*a \ \& \ Ba/H \ \& \ K)$ is the probability that an object a is randomly chosen from the class of ravens and is black, given $H \ \& \ K$, and this probability is surely not 1. In the second place, comparing $\Pr(\neg B^*b \ \& \ \neg Rb/K)$ and $\Pr(R^*a \ \& \ Ba/K)$ involves a comparison of the probability that an object will be randomly sampled from the class of ravens with the probability that it will be randomly sampled from the class of nonblack things, and such a comparison seems peripheral to the paradox at best.

Horwich's basic idea can be brought to fruition by putting into the background knowledge $\hat{K}$ the information that R^*a and $\neg B^*b$. Bayes's theorem can then be legitimately applied to the new $\hat{K}$ to conclude that

$$\Pr(H/Ra \ \& \ Ba \ \& \ \hat{K}) = \Pr(H/\hat{K})/\Pr(Ba/\hat{K}) \tag{3.7}$$

and

$$\Pr(H/\neg Rb \ \& \ \neg Rb \ \& \ \hat{K}) = \Pr(H/\hat{K})/\Pr(\neg Rb/\hat{K}). \tag{3.8}$$

Thus, relative to this $\hat{K}$, the evidence $Ra \ \& \ Ba$ has more confirmational value vis-à-vis the ravens hypothesis than does $\neg Rb \ \& \ \neg Bb$ just in case $\Pr(\neg Rb/\hat{K}) > \Pr(Ba/\hat{K})$. A further application of the principle of total probability shows that this latter inequality holds just in case $\Pr(\neg Ba/\neg H \ \& \ \hat{K}) > \Pr(Rb/\neg H \ \& \ \hat{K})$. This last inequality presumably does hold in our universe, for given that some ravens are nonblack ($\neg H$), we are more likely to produce one of them by sampling from the class of ravens than by sampling from the class of nonblack things simply because of the known size and heterogeneity of the class of nonblack things as compared with the known size of the class of ravens. Suppes is thus vindicated after all, since the greater confirmatory power of $Ra \ \& \ Ba$ over $\neg Rb \ \& \ \neg Bb$ has to do with the relative threats of falsification. In this way Bayesianism pays a backhanded compliment to Popper's methodology; namely, it is precisely because, contrary to Popper, inductivism is possible that the virtues of sincere attempts to falsify can be recognized.[11]

Similar points are made by Gaifman (1979), although his assumed sampling procedure is somewhat different. Let $\tilde{K}$ report that c was drawn at random from the universe and found to be a raven and that d was also drawn at random from the universe and found to be nonblack. An analysis like the one above shows that

$$\Pr(H/Rc \& Bc \& \tilde{K}) > \Pr(H/\neg Rd \& \neg Bd \& \tilde{K})$$

just in case $\Pr(\neg Bc/\neg H \& \tilde{K}) > \Pr(Rd/\neg H \& \tilde{K})$.

But the procedure of sampling from the universe at large can be wasteful, since it can produce relatively useless results, such as $\neg Re \& Be$. Moreover, one can wonder whether the evidence $Ra \& Ba$, under the assumption that a was drawn at random from the class of ravens, gives better confirmational value than the evidence $Rc \& Bc$, under the assumption that c was drawn at random from the universe at large, i.e., whether

$$\Pr(H/Ra \& Ba \& \hat{K} \& \tilde{K}) > \Pr(H/Rc \& Bc \& \hat{K} \& \tilde{K}).$$

I leave it to the reader to ponder this question with the clue that the answer is positive just in case

$$\Pr(\neg Ba/\neg H \& \hat{K} \& \tilde{K}) > \Pr(\neg Bc/\neg H \& \hat{K} \& \tilde{K}).^{12}$$

4 Bootstrapping and Relevance Relations

In *Theory and Evidence* (1980) Glymour saw bootstrapping relations not only as a means of extending Hempel's instance confirmation to theoretical hypotheses but also as an antidote to Duhem and Quine holism. It makes a nice sound when it rolls off the tongue to say that our claims about the physical world face the tribunal of experience not individually but only as a corporate body. But scientists, no less than business executives, do not typically act as if they are at a loss as to how to distribute praise through the corporate body when the tribunal says yea, or blame when the tribunal says nay. This is not to say that there is always a single correct way to make the distribution, but it is to say that in many cases there are firm intuitions. Bootstrap relations would help to explain these intuitions if they helped to explain why it is that for some but not all H's that are part of a theory T, E bootstrap-confirms H relative to T.

As a sometime Bayesian I now think that bootstrapping should be abandoned in favor of a Bayesian analysis. Bayesians can be sympathetic to the two motivations for bootstrapping mentioned above in section 2. At the same time Bayesians can recognize that any account of confirmation modeled on Hempel's approach will have two fatal flaws. (1) For Hempel, whether or not E confirms H depends only on the syntax of E and H. But from Goodman we know this to be wrong (see section 2 above and chapter 4). (2) For Hempel, confirmation is a two-place relation. But from the ravens paradox and other examples we know that background information K must be brought into the analysis to get an illuminating treatment. The relevance of these points to bootstrapping can be brought into focus with the help of Christensen's (1983) examples.

Let T have as its axioms H_1: $(\forall x)(Rx \to Bx)$, and H_2: $(\forall x)(Rx \to Hx)$, the former of which is our old friend the ravens hypothesis and the latter of which asserts that all ravens live a happy afterlife in bird heaven. At first blush, evidence from the observation of the color of a raven is directly relevant to H_1 but is irrelevant to H_2, even relative to T. But Christensen shows how, with a little logical flimflam, such evidence leads to a bootstrap confirmation of H_2 relative to T. On the standard conception of theories, T is the logical closure of $\{H_1, H_2\}$. Thus it is part of T that

H_3: $(\forall x)[Rx \to (Bx \leftrightarrow Hx)]$.

From E: $Ra \& Ba$, we can deduce via H_3 that $Ra \& Ha$, which is a Hempel positive instance of H_2. Moreover, the possible alternative evidence E': $Ra \& \neg Ba$, leads via H_3 to $Ra \& \neg Ha$, which is a counterinstance of H_2. Together these "computations" constitute a positive bootstrap test of H_2. But intuitively, only phony-baloney confirmation/testing has taken place. A revised set of bootstrap conditions proposed by Glymour (1983) rule out this particular example, but Christensen (1990) has shown how the counterexamples can be revived in a more complicated form.[13]

One could seek further restrictions to rule out the new counterexamples, but this now seems to me to be a mistake—we do not want a once-and-for-all answer to, Does E confirm H relative to T? that is independent of the interpretation of the nonlogical constants in E, H, and T and also independent of the background knowledge.

To make the point more concrete, let me use another of Christensen's examples, which is structurally identical to the above ravens case. Now T is the logical closure of H_1: $(\forall x)(Sx \to Ax)$, and H_2: $(\forall x)(Sx \to Vx)$. H_1 is intended to assert that anyone with certain disease symptoms has the antibodies to a certain virus, while H_2 is intended to assert that anyone with the said symptoms has been infected by the said virus. T contains

H_3: $(\forall x)[Sx \rightarrow (Ax \leftrightarrow Vx)]$. The evidence E: Sa & Aa, leads via H_3 to a positive instance of H_2, while the alternative possible evidence E': Sa & $\neg Aa$, leads via H_3 to a negative instance of H_2. Although structurally identical to the former example, we are not so ready to see phony-baloney confirmation/testing here.

To diagnose the felt asymmetries between the two cases, we need to know to what end the three-place Glymourian relation 'E bootstrap-confirms/tests H relative to T' is to be put. E bootstrap-confirms H relative to T cannot be taken to imply that, assuming T to be true or well confirmed, E confirms H, for in the cases at issue H is part of T. Rather, the most plausible usage is in adjudicating questions of evidential relevance. Note that in these examples Hempel's version of the "prediction criterion" of confirmation is satisfied; i.e., E is of the form E_1 & E_2, where $\{T, E_1\} \models E_2$ but $E_1 \not\models E_2$, while E' is of the form E'_1 & E'_2, where $\{T, E_1\} \models \neg E'_2$ but $E'_1 \not\models \neg E_2$. The antiholist then asks, if E is found to hold, to which parts of T can the praise for the successful prediction be attributed? If E' is found to hold, on which parts of T can the blame for the unsuccessful prediction be laid?

With this interpretation of bootstrapping, the Bayesian diagnosis of the counterexamples is straightforward. H gets praise from E if, relative to K, E incrementally confirms H, and H gets blame from E' if, relative to K, H is incrementally disconfirmed by E'. The bird-heaven case gave off a bad odor since our current background knowledge K would have to be radically altered for Ra & Ba to incrementally confirm H_2 or for Ra & $\neg Ba$ to incrementally disconfirm H_2. Indeed, given the tenets of traditional empiricism, we could never get to an alternative K where this would happen. By contrast, the virus case smelled sweeter even though, from the point of view of bootstrapping, it is structurally identical to the bird-heaven case. A possible reason is that in the virus case H_2 will get praise from Sa & Aa and blame from Sa & $\neg Aa$ if K makes likely the proposition that all and only those people who have been infected by the virus have antibodies to it, a not implausible situation.

It might be complained that while such a diagnosis does in fact help to explain intuitions, it is irrelevant to the original project; the aim of that project was to provide an internalist analysis of relevance relations, and given that aim, it is illegitimate to bring in K. The response to this complaint parallels the response to Hempel's complaint that background information about the relative sizes of the classes of ravens and nonblack things is irrelevant to his project, which concerns only the two-place relation '*E* confirms *H*'; namely, no interesting account of confirmation can be developed if *K* is left out of the picture.

Aron Edidin (1988) has maintained that the core of the program of relative confirmation is left untouched by Christensen's examples. I think that there is a sense in which Edidin's contention is correct, but by the same token I think that the program of relative confirmation can be seen to be drained of much of its interest. Let us suppose that the core of the program is concerned with the relation '*E* confirms *H* relative to auxiliaries *A*', where typically the auxiliaries do not include *H* itself. Edidin's point is that there is nothing in Christensen's examples to suggest that the apparatus developed in *Theory and Evidence* is not adequate to provide a correct explication of this relation. Thus in Christensen's ravens example there is nothing counterintuitive to maintaining that *E*: *Ra & Ba*, does confirm H_2: $(\forall x)(Rx \rightarrow Hx)$, relative to the auxiliary assumption H_3: $(\forall x)[Rx \rightarrow (Bx \leftrightarrow Hx)]$. This seems to me correct in the following respect: in the sense in which Hempel could say that *E*: *Ra & Ba*, confirms H_1: $(\forall x)(Rx \rightarrow Bx)$, it is also natural by extension to say that *E* confirms H_2 relative to H_3.

But if the core of the program of relative confirmation is left untouched, it remains to ask what purpose is served by the program. Two responses suggest themselves. First, we can hope to use the relation '*E* confirms *H* relative to *A*' to explicate theory-relative confirmation. Thus, we can say that '*E* confirms *H* relative to *T*', where *T* typically contains *H*, means that there is an appropriate *A* in *T* such that *E* confirms *H* relative to *A*. Here the appropriateness of *A* is supposed to guarantee that the resulting confirmation/disconfirmation of *H* relative to *T* by *E* implies that the praise/blame for *T*'s passing/failing to pass an HD test can be attached to *H*. The presumption of *Theory and Evidence* was that the appropriateness of *A* can be settled purely in terms of structural relations among *A*, *H*, *E*, and *T*. This presumption is belied by the analysis above of Christensen's examples, which shows that the parceling out of praise and blame depends on the epistemic status of *A*, which in turn depends upon the background knowledge.

The second response is that getting a handle on relative confirmation is useful in deciding how evidence affects the credibility of hypotheses and in turn the credibility of theories of which the hypotheses are parts. But again, the epistemic status of the auxiliaries must be taken into account. Edidin's discussion indicates that the move from '*E* confirms *H* relative to *A*' to '*E* contributes to the credibility of H' is a tricky one; it requires not only that the auxiliaries A "must themselves be credible." In some cases it requires also that "their credibility must be substantially independent of the credibility of the evidence" (p. 268) and in other cases that they have "antecedent credibility independent of that of the hypotheses" (p. 269). But what exactly do these requirements come to? I submit that no precise answer can be given without invoking the Bayesian apparatus. Further, the answer this apparatus yields is that no answer can be given in the abstract: it depends on the background information K, and it depends not just on the logico-structural relations involved in the HD and bootstrapping account of relative confirmation but also on the intended interpretation of the nonlogical terms in E, H, and A.

The complaint here is not that, on pain of circularity, HD or bootstrapping relations of relative confirmation cannot figure in an account of how evidence bears on the credibility of theoretical hypotheses; rather, the complaint is that such relations may not contribute in any perspicuous way to the assessment of that bearing. Consider again the simpler case of the confirmation of observational hypotheses. How, for example, does evidence about the color of ravens and nonravens bear on the credibility of the hypothesis that all ravens are black? By now, I hope, the reader is convinced that an illuminating path to an answer need not take the form of first deciding when E Hempel-confirms H and then trying to puzzle out the further conditions necessary for the move from Hempel-confirmation to an incremental increase in credibility. The moral here has double strength when we move from Hempel-confirmation of observational hypotheses to the more complicated case of relative confirmation of theoretical hypotheses.

5   Variety of Evidence and the Limited Variety of Nature

It is a truism of scientific methodology that variety of evidence can be as important or even more important than sheer amount of evidence. An adequate account of confirmation is not under obligation to give an unqualified endorsement to all such truisms, but it should be able to identify the valid rationale (if any) of such truisms.

A Bayesian explanation of the virtue of variety of evidence would concentrate on the ability of variety to contribute to a significant boost in the posterior probability of a hypothesis. To illustrate how part of the explanation might go, consider again the HD case where $H, K \models E$, and suppose that E is E_1 & E_2 & ... & E_n, where the E_i report the outcomes of performing some one experiment over and over or alternatively the outcomes of a series of different experiments. The most helpful form of Bayes's theorem to cover this situation is

$\Pr(H/E$ & $K)$

$$= \frac{\Pr(H/K)}{\Pr(E_1/K) \times \Pr(E_2/E_1 \ \& \ K) \times \cdots \times \Pr(E_n/E_1 \ \& \ ... \ \& \ E_{n-1} \ \& \ K)}.$$
(3.9)

As we will see in chapter 4, if $\Pr(H/K) > 0$, the factor $\Pr(E_n/E_1$ & ... & E_{n-1} & $K)$ must go to 1 as n grows without bound. This factor gives the probability of the next experimental outcome predicted by H, conditional on the background information K and the information that the previous predictions have been borne out. The more slowly this probability approaches 1, the smaller the denominator (for a given n) and hence the larger the posterior probability of H (for a given n). This is exactly where variety of evidence enters, for the more various the experiments, the slower one would expect the approach to certainty to be for the next outcome.[14] At one extreme is the case where the E_i are the outcomes of repeating the same experiment consisting, say, of measuring over and over again a quantity believed to have a stable value. Then with appropriate assumptions K about the reliability of the measuring apparatus, only a few repetitions are needed to achieve near certainty for the next instance, and amassing a large number of further instances achieves little gain for the posterior probability of H. At the other extreme is the case where the E_i are the outcomes of experiments that are not only different but seem quite unrelated. Then new instances will make for a bigger gain in the posterior probability of H.[15]

These remarks have value only if we already have a grip on the notion of variety of evidence. But rather than trying to give an independent analysis of variety, what I would like to suggest is that the observations above can be given a new twist and used to define 'variety of evidence' through rate of increase in the factors

$\Pr(E_n/E_1$ & ... & E_{n-1} & $K)$.[16]

Such an analysis has two consequences, one of which is obvious, the other of which is a little surprising.

The obvious consequence is that the notion of variety of evidence has to be relativized to the background assumptions K, but there is no more than good scientific common sense here, since, for example, before the scientific revolution the motions of the celestial bodies seemed to belong to a different variety than the motions of terrestrial projectiles, whereas after Newton they seem like peas in a pod.

The less obvious consequence is that induction, or a necessary condition for it, presupposes a limited variety in nature, as Keynes (1962) tried to teach us. As already remarked, $Pr(H/K) > 0$, which is necessary for the probabilification of H, implies that

$$Pr(E_n/E_1 \& \ldots \& E_{n-1} \& K) \to 1$$

as $n \to \infty$. This means that from the point of view of the proposed analysis of variety, E_n for large enough n cannot be counted as various with respect to $E_1, E_2, \ldots, E_{n-1}$, contrary to what our untutored intuitions might have told us. The fact that the E_i are unified in the very minimal sense of being entailed by a single H to which we assign a nonzero prior eventually forces us to see them as nonvarious.

Another aspect of the importance of variety of evidence arises in conjunction with eliminative induction, whose virtues are touted in chapter 7. Bayes's theorem in the form (2.2) shows how the probability of a hypothesis is boosted by evidence that eliminates rival hypotheses. Thus variety of evidence can be analyzed from the point of view of how likely the evidence is to produce efficient elimination.[17]

6   Putnam and Hempel on the Indispensability of Theories

Induction by enumeration is inadequate for capturing many of the inferences routinely made in the advanced sciences, as is brought out very nicely by the following example of Putnam's (1963a). Imagine that you were a member of the Los Alamos Project during World War II. As you prepare for the first test of what you hope will be an atomic bomb, you consider prediction H: when these two subcritical masses of U_{235} are slammed together to form a supercritical mass, there will be an atomic explosion. H has a counterpart in purely observational terms, namely H': when these two rocks are slammed together, there will be a big bang. If E is the sum of the directly relevant observations made up to this juncture, there is no way for an inductivist who limits himself to simple enumeration to move from E to a confidence in H'. For up to now there have been no recorded cases of rocks of this kind exploding, but there have been many recorded cases of rocks of this kind being slammed together without exploding (because critical mass was never reached). Nevertheless, you and your fellow project scientists are confident of H'. Why?

The Bayesian is happy to supply the answer. You were in possession of a theory T of the atomic nucleus that entails H'. Applying the principle of total probability to the total available observational evidence $E \& \hat{E}$ gives

$$\Pr(H'/E \& \hat{E}) = \Pr(T/E \& \hat{E}) + \Pr(H'/\neg T \& E \& \hat{E}) \times \Pr(\neg T/E \& \hat{E}).$$

Thus if your opinions conformed to the probability calculus, your confidence in H' should have been at least as great as your confidence in T. And the combination of E and $\hat{E}$ made you somewhat confident of T (because, for example, T entails other experimental regularities whose positive instances are recorded by $\hat{E}$). Further, $\neg T$ includes other theories that also entail H' or make H' highly probable, and $E \& \hat{E}$ made you somewhat confident of those theories. The upshot was that you were more than somewhat confident of H'.

Putnam used this story to register a complaint against any explication of degree of confirmation that makes the confirmation of H' on $E \& \hat{E}$ independent of the presence or absence in the language of predicates not occurring in H', E, or $\hat{E}$ (what Carnap in 1950 and 1952 called an inductive method of the "first kind"). In terms of the present example, such an explication implies that $\Pr(H'/E \& \hat{E})$ can be assessed in a language that contains only observational predicates. But since expressions involving T cannot occur in such a language, the explanation above of the expectations of the Los Alamos scientists cannot be stated in such a setting. To provide an explanation within the strictures of an inductive method of the first kind, it must be supposed that the scientists involved would have had the same degree of confidence in H' had they never considered T, a highly implausible supposition, to say the least. Of course, it could be replied that the failure of inductive methods of the first kind to accord with the actual psychology of scientists may be ignored, since the task of explicating degree of confirmation is a normative rather than a descriptive one. The rejoinder is that the normative status of a proposed explication comes into question when the explication fails to accord with what the history of science provides as paradigm cases of good inferences. In effect, Carnap agreed with this rejoinder in his response to Putnam. He wrote that for situations of this kind we must construct a new inductive logic which refers to the theoretical language instead of the observational language. I would say that the scientists at the time in question would indeed have been willing to bet on the positive success of the first nuclear explosion on the basis of the available evidence, including results of the relevant laboratory experiments. Inductive logic must reconstruct this willingness by ascribing to $c(H, E)$ a considerable positive value.[18] (1963b, p. 988)

Is there an argument here for scientific realism? Not much of one, but something is better than nothing. Consider the position of an antirealist who is neither an instrumentalist nor an inductive skeptic with respect to observational predictions but who is an inductive skeptic with respect to theoretical claims. In the Los Alamos example such an antirealist will agree that reasonable expectations about the explosion prediction H' can be formed on the basis of E & $\hat{E}$. He also agrees that the nuclear theory T has a truth value and that the proposition asserting that T is true is not merely a disguised way of asserting that observational predictions of T are correct. But he nevertheless denies that the observational evidence E & $\hat{E}$ serves as a basis for a reasonable belief in the truth of T. Such an antirealist is very much in the same position as someone who uses a Carnapian method of the first kind, and whatever objections can be brought against the latter can also be brought against the former.

The above considerations also help to illuminate Hempel's (1958) proposed resolution of the "theoretician's dilemma." On Hempel's formulation, the dilemma runs thus: either theoretical terms fulfill their function of systematizing deductive connections among observation statements or they don't. If they don't, they are obviously dispensable. If they do, they are likewise dispensable, since Craig's (1956) lemma shows that the observational consequences of an axiomatizable theory can always be reaxiomatized in purely observational vocabulary. Hence theoretical terms are dispensable. Hempel's response was that theories may be indispensable because they serve to establish *inductive* as well as *deductive* connections.

T might be said to be essential to establishing inductive connections among observables if there are observation sentences O_1 and O_2 such that $\Pr(O_2/T \& O_1) > \Pr(O_2/O_1)$, or more interestingly, if $\Pr(O_2/T \& O_1) > \Pr(O_2/O_T \& O_1)$, where O_T is a sentence logically equivalent to the set of observational consequences of T.[19] The first condition is certainly satisfied in the Los Alamos example with $O_1 = E$ & $\hat{E}$ and $O_2 = H'$, and for sake of argument we may suppose that the second condition is satisfied as well.

But on further reflection, these facts do not by themselves establish the claimed indispensability of T. In the Los Alamos example, the key question is what degree of confidence to put in H' on the basis of the total available evidence E & $\hat{E}$. Thus in this example the claim that theories are indispensable for purposes of inductive systematization must be understood as the claim that the evaluation of $\Pr(H'/E$ & $\hat{E})$ depends in some essential way on T. But what way is this? I suggest that the answer must be the one supplied by my discussion of Putnam's story. And I would further suggest that the moral of the story can be generalized.

Suppose that for purposes of scientific investigation of a certain domain, an inductive agent adopts a language $\mathscr{L}$ and a degree-of-belief function Pr on the propositions $\mathscr{A}$ of $\mathscr{L}$. We may suppose that $\mathscr{L}$ is a purely observational language. Subsequently the agent expands her language to $\mathscr{L}'$, which includes theoretical predicates, and adopts a degree-of-belief function Pr' for the propositions $\mathscr{A}' \supset \mathscr{A}$ of the new language. Even though she is a rational agent, it may very well be that Pr' restricted to $\mathscr{A}$ does not coincide with her previous belief function Pr. Of course, this phenomenon has nothing to do per se with the observational/theoretical distinction; it is merely a corollary of the point that the probability assigned to a proposition may depend upon the possibility set in which the proposition is imbedded. The moral here has an intralanguage counterpart. Within, say, the language of physics as it is constituted at any particular time, physicists are explicitly aware of only a small portion of the possible theories that can be formulated in the language. When new theories are formulated, the range of the explicitly recognized possibilities being thereby expanded, the probabilities of previously considered hypotheses and theories may change. This matter is taken up in chapters 5 and 7.

A striking consequence emerges when we combine such morals with Carnap's principle of tolerance, according to which "everyone is free to use the language most suited to his purpose" (1963a, p. 18). Since the exercise of this freedom is guided to a large extent by pragmatic factors, and since degree of confirmation is affected by the choice of language, the implication is that evidential support has a pragmatic dimension. Pure personalists will hardly be shocked by this consequence, but those who want confirmation theory to deliver rational and objective degrees of belief may not be so shock-proof.

Those who do find such a consequence repugnant may want to consider restrictions on the principle of tolerance, but it is hard to see how a principled intolerance is to be implemented. Alternatively, the consequence can be avoided by doing confirmation theory in a universal language adequate for reconstructing all past and future scientific endeavors. But even if such a utopian scheme is possible, its relevance to the actual practice of science, which takes place in a context far from utopia, is tenuous.[20] Rather than try to avoid the consequence, I recommend a cautious embrace. Chapter 7 gives a concrete example of one form the embrace might take.

7 The Quine and Duhem Problem

If hypotheticodeductivism were the only tool available for assessing evidence, we would be at a loss in making judgments about how evidence bears differentially on the components of a scientific theory. Some additional tool is thus sorely needed. In section 4, I found fault with Glymour's attempt to parcel out praise and blame using bootstrapping relations, and I intimated that the parceling out is best accomplished with Bayesian means. Sometimes a Bayesian analysis supports a kind of holism. Thus if T consists of the conjunction of T_1 and T_2, and if T contradicts $E \& K$, the blame may attach to T as a whole without sticking to either component T_1 or T_2. Indeed, Wesley Salmon (1973) has provided an example where, relative to K, E incrementally confirms each of T_1 and T_2, even though T is refuted by $E \& K$.[21] In more typical cases of refutation, however, our intuitions suggest that the blame does stick to one or another component of the theory and also that it sticks more firmly to some components than to others.

   An example of how the Bayesian apparatus can be used to support such intuitions in historically realistic cases has been given by Jon Dorling (1979). Suppose that theory T consists of core hypotheses T_1 and auxiliary assumptions T_2; that $T_1 \& T_2 \models E'$; and finally that nature pronounces E, which is incompatible with E'.[22] Dorling assumes that T_1 is probabilistically irrelevant to T_2 (that is, $\Pr(T_2/T_1) = \Pr(T_2)$), that the priors $\Pr(T_1) = k_1$ and $\Pr(T_2) = k_2$ satisfy $k_1 > k_2$ and $k_1 > .5$, while the likelihoods $\Pr(E/\neg T_1 \& T_2) = k_3$, $\Pr(E/T_1 \& \neg T_2) = k_4$, and $\Pr(E/\neg T_1 \& \neg T_2) = k_5$ satisfy $k_3 \ll k_4, k_5 \ll 1$. Then Bayes's theorem shows that the blame falls more heavily on the auxiliaries T_2 than on the core T_1. If we take the time to be the mid nineteenth century, T_1 to be Newton's theory of motion and gravitation, T_2 the assumption that tidal effects do not influence lunar secular acceleration, and E the observed secular acceleration of the moon, then Dorling argues that plausible values of the relevant probabilities are $k_1 = .9$, $k_2 = .6$, $k_3 = .001$, $k_4 = k_5 = .05$. With these values he finds that $Pr(T_1/E) = .8976$ and $Pr(T_2/E) = .003$, so that the refuting evidence E only slightly reduces the probability of the core of the theory, while strongly undermining the auxiliary.[23]

Assuming that Dorling's reconstruction of the prevailing degrees of belief is historically correct, we are presented with a Bayesian success story in the form of an explanation of the attitudes and behavior displayed by the scientific community during an important incident in nineteenth-century astronomy. But what we don't yet have is a solution to the Quine and Duhem problem, at least not if what we demand of a solution is a demonstration that one way of parceling out the blame is rationally justified while others are not. For it is perfectly compatible with Bayesian personalism to assign values to k_1 through k_5 that make T_1 the goat while rendering T_2 blameless.[24] We have arrived at one aspect of the general problem of the objectivity of scientific inference, a problem that will occupy us from chapter 6 onward. I will note in advance that while much of the attention on the Bayesian version of this problem has focused on the assignments of prior probabilities, the assignments of likelihoods involves equally daunting difficulties.

In the present context the difficulties can be illustrated by noting that when T_1 & $T_2 \vdash \neg E$ but nature pronounces E, then blame attaches squarely to T_1 in the sense that $Pr(T_1/E) \ll Pr(T_1)$ just in case

$$Pr(E/T_1 \& \neg T_2) \times Pr(\neg T_2/T_1) \ll Pr(E/\neg T_1 \& T_2) \times Pr(T_2/\neg T_1)$$

$$+ Pr(E/\neg T_1 \& \neg T_2) \times Pr(\neg T_2/\neg T_1).$$

In general, none of the factors involved has an objective character, and a large variability can be expected in the values assigned by different persons. Dorling's argument that this inequality fails in his historical case study is based on the assumption that $Pr(E/\neg T_1 \& T_2)$ is small—an assumption Dorling takes to be justified because (he says) no plausible rival to Newton's theory could predict E either quantitatively or qualitatively. This justification succeeds if $\neg T_1$ is limited to rivals actually constructed by nineteenth-century physicists. But a critic of this analysis might well ask why pronouncements about what it is and isn't rational to believe in the face of E should depend on the vicissitudes of which of the myriad possible theories happened to be constructed by physicists of the time.

Let us attempt to add some objectivity by moving to a simple if unrealistic case. Assume first that T_1 and T_2 are probabilistically irrelevant to one another. Assume second that we can parse $\neg T_1$ as $T_1^1 \vee T_1^2 \vee \ldots \vee T_1^n$, where the T_1^i are pairwise inconsistent, and that we can parse $\neg T_2$ as $T_2^1 \vee T_2^2 \vee \ldots \vee T_2^m$, where the T_2^j are also pairwise inconsistent. Assume finally that T_1 or any one of the T_1^i when conjoined with T_2 or any one of the T_2^k together entail a definite prediction for the phenomenon in question. Then the condition for blame to attach to T_1 becomes

$$\sum_j \Pr(T_2^j) \ll \left[\frac{\Pr(T_2)}{\Pr(\neg T_1)} \right] \times \sum_i \Pr(T_1^i) + \left[\frac{\Pr(\neg T_2)}{\Pr(\neg T_1)} \right] \times \sum_k \Pr(T_1^k),$$

where the sum on j is taken over values such that $T_1 \ \& \ T_2^j \vdash E$, the sum on i is taken over values such that $T_1^i \ \& \ T_2 \vdash E$, and the sum on k is taken over values such that $T_1^k \ \& \ \neg T_2 \vdash E$ (i.e., $T_1^k \ \& \ T_2^j \vdash E$ for every value of j). At first this result is a little disconcerting, since in an effort to objectify the problem, we have reduced it to one involving judgments of priors. What we can hope is that the priors used in this context are posteriors taken from another context and that the latter have been objectified through the weight of accumulated evidence.

The result of accumulating evidence has been investigated by Redhead (1980) under a different set of assumptions. He invites us to consider a series of refutations of the core (T_1) plus auxiliary (T_2). T_2 is replaced by T_2' to accommodate the evidence E refuting $T_1 \ \& \ T_2$; then new data F that refutes $T_1 \ \& \ T_2'$ is found; T_2' is replaced by T_2'' to accommodate F; etc. If each of the successive auxiliaries is given an initial weight of .5, and if the likelihoods of each new piece of evidence (given $\neg T_1 \ \& \ T_2' \cdots$, $T_1 \ \& \ \neg T_2' \cdots$, or $\neg T_1 \ \& \ \neg T_2' \cdots$) are equal and substantially less than 1, then the probability of T_1 is quickly driven down toward 0 by the series of refutations. This is an interesting result, but it does not provide a resolution of the original problem.

The upshot is that we have a highly qualified success for Bayesianism: the apparatus provides for an illuminating representation of the Quine and Duhem problem, but a satisfying solution turns on a solution to the general problem of objectivity of scientific inference, a matter that will occupy us in coming chapters.

8   Conclusion

The reader does not have to share the details of the sentiments I have expressed above to be convinced that applying the Bayesian apparatus to topics like the paradox of the ravens, the variety of evidence, the role of theories in scientific inference, and the problem of Quine and Duhem leads to fruitful avenues of investigation. There are many more examples of fruitfulness that could be given. Some will be developed in chapter 4 in the context of responses to challenges to Bayesianism confirmation theory. Others can be found in such Bayesian tracts as Rosenkrantz 1981, Horwich 1982, and Howson and Urbach 1989. Franklin (1986, 1990) supplies excellent case studies of experiments in physics and makes an attempt to provide a Bayesian rationale for the strategies he sees experimental physicists using to validate their results.

4 Challenges Met

Despite or perhaps because of its successes, Bayesianism is not without its detractors. One of the most serious charges against it is that its machinery does not apply to the confirmation of universal hypotheses about an infinity of individuals, since (the charge goes) the prior and thus the posterior probability of such a generalization will be flatly 0. Three versions of this worry are examined in sections 1 to 3. Section 4 explores a different worry expressed by Karl Popper and David Miller. They argue that even when the probabilification of a hypothesis takes place, no genuine inductive support can be seen in the incremental boost in probability. Section 5 is devoted to Richard Miller's charge that Bayesianism is just as broken-backed as is HD methodology because the notorious problem of adhocing the auxiliary hypotheses that besets the latter has analogues that vitiate the former. Section 6 takes up Grünbaum's worry that Bayesianism commits its practitioners to an unbridled and implausible form of instantian inductivism. Section 7 explores the ability of Bayesianism to cope with Goodman's "new problem of induction." Finally, section 8 asks whether Bayesianism can account for the importance of novel predictions.

1 The Problem of Zero Priors: Carnap's Version

A *Carnapian confirmation function* $c(H, E)$ for a language is a conditional probability function (see appendix 1 of chapter 2) defined on pairs of sentences H, E of the language, where E is noncontradictory. From the axioms of conditional probability it follows that $c(H \& E, t) = c(E, t) \times c(H, E)$, where t is a tautology. If c is strictly coherent (see chapter 2) so that $c(E, t) \neq 0$ for a noncontradictory E, then we can write $c(H, E) = c(H \& E, t)/c(E, t)$. If we set $m(\cdot) \equiv c(\cdot, t)$, $c(H, E) = m(H \& E)/m(E)$, we see that the confirmation function is determined by the *measure function m*. (When c is not strictly coherent, the story becomes more complicated, but the details will not be rehearsed here.)

Following Carnap (1950, 1952), let us now specialize to a language L_N^K containing K monadic predicates $P_1, P_2, \ldots, P_K$, assumed to be logically independent, and N individual constants $a_1, a_2, \ldots, a_N$. A *state description* specifies for each P_i and a_j whether or not P_i applies or fails to apply to a_j. In this setting, a measure function is an assignment to the state descriptions of positive weights that sum to 1, and the $c(H, E)$ determined by this measure function is the ratio of the sum of the weights attached to the state descriptions in which both H and E hold to the sum of the weights of the state descriptions in which E holds.

If the only significance of the subscripts on the a_j's is that different subscripts indicate that the individual constants denote different individuals, then it is natural to require the following:

R1   The c functions are *symmetric*, i.e. $c(H, E) = c(\Pi(H), \Pi(E))$.

Here $\Pi(X)$ indicates the result of replacing the individual constants in X by their counterparts from an arbitrary permutation of the individual constants. For some time Carnap was enamored of a particular symmetric c function, c^*, which is defined by assigning equal measures to each *structure description* and to each state description within a structure description, where a structure description is a maximal set of state descriptions each of which can be transformed into any other by a permutation of the individual constants. Carnap's assumptions here can be seen as a generalization of Bayes's notion that when only two outcomes are possible, the prior probability of any given number of successes in n trials is $1/(n + 1)$ (see chapter 1).

In *The Continuum of Inductive Methods* (1952) he saw the need for a more general approach, which I will now briefly sketch. By a *Q-property*, let us understand the conjunctive property formed by choosing for each $i = 1, 2, \ldots, K$ either P_i or $\neg P_i$. There are $\kappa = 2^K$ Q-properties of L_N^K. These properties are mutually exclusive and exhaustive, so we may think of them as boxes that partition the universe. Consider the evidence E^n that tells us which of the n individuals examined so far fall into which Q boxes. If H_k^{n+1} is the hypothesis that the next individual examined will fall into the kth Q box, then (R1) implies that $c(H_k^{n+1}, E^n)$ is a function purely of n, κ, and the box occupation numbers $n_1, n_2, \ldots, n_\kappa$ ($\sum_{i=1}^{\kappa} n_i = n$). In *The Continuum of Inductive Methods* Carnap required something much stronger:

R2   The value of the confirmation function $c(H_k^{n+1}, E^n)$ depends on n_k but not on the other occupation numbers.

The problem now is to determine the general form of $c(\cdot, \cdot)$ satisfying these constraints. Apparently unaware that W. E. Johnson (1932) had successfully tackled the same problem many years before, Carnap showed by a very ingenious argument that (when $\kappa > 2$) any c function obeying (R2) can be written as

$$c(H_k^{n+1}, E^n) = \frac{n_k + \lambda/\kappa}{n + \lambda}, \tag{4.1}$$

where λ lies in the interval $[0, +\infty]$. The parameter λ can be viewed as an index of inductive caution that determines how fast we are willing to learn from experience. The value $\lambda = 0$ (which is forbidden by strict coherence) corresponds to the "straight rule" of induction: $c(H_k^{n+1}, E^n) = n_k/n$. At the other extreme, $\lambda = \infty$ implies that there is no learning from experience in that the probability that the next individual falls into the kth Q box is $1/\kappa$, regardless of how many of the objects examined so far have fallen into this box. Setting $\lambda = \kappa$ gives us back c^*. Given the special values fixed by (4.1), it is easy to show that the value of $c(H, E)$ for any sentences H and E of L_N^K is uniquely determined.

This scheme cannot be extended in any natural way so as to permit the confirmation of universal hypotheses in a universe containing a countable infinity of individuals. For from (4.1) it follows that $c(\&_{j \leqslant n} P_i a_j, t)$ tends to 0 as n tends to infinity. Since $(\forall j) P_i a_j \models \&_{j \leqslant n} P_i a_j$ for any n, $c((\forall j) P_i a_j, t) \leqslant c(\&_{j \leqslant n} P_i a_j, t)$, with the upshot that the universal generalization must receive a c value of 0 in an infinite universe.

Kemeny (1963) was so disturbed by this result that he flirted with the idea that confirmation functions have to be constructed from measures that take real-valued functions rather than real numbers as values. However, nothing this drastic is required to permit the confirmation of universal generalizations in infinite universes. Hintikka (1966) and Hintikka and Niiniluoto (1980) retain (R1) but modify (R2) to allow that $c(H_k^{n+1}, E^n)$ may depend upon the total number of boxes occupied in addition to n_k, n, and κ. They show that in their more liberal systems the prior probability of a universal generalization can be nonzero in an infinite universe and that the posterior probability of the universal generalization can approach 1 as the positive instances accumulate.

We need not go into the details of the Hintikka and Niiniluoto systems to understand how the symmetry requirement (R1) and its generalizations are compatible with the confirmation of universal generalizations. For simplicity, consider a monadic predicate 'P' and an infinite sequence of individual constants $a_1, a_2, \ldots$. And consider a probability measure Pr defined on all the Pa_n and on all finite truth functional compounds of such atomic formulas. In concert with (R1), we suppose that Pr is *exchangeable* (de Finetti), meaning that for any k,

$$\mathrm{Pr}(\pm Pa_{n_1} \& \pm Pa_{n_2} \& \ldots \& \pm Pa_{n_k})$$

$$= \mathrm{Pr}(\pm Pa_{n_1'} \& \pm Pa_{n_2'} \& \ldots \& \pm Pa_{n_k'}), \tag{R1$'$}$$

where the $\pm$ indicates that either the predicate or its negation may be chosen, and $n'_1, n'_2, \ldots, n'_k$ stands for an arbitrary permutation of $n_1, n_2, \ldots, n_k$. De Finetti's representation theorem then shows that the Pr probability that k out of n individuals will be P is

$$\binom{n}{k} \int_0^1 \theta^k (1 - \theta)^{n-k} \mu(d\theta) \tag{4.2}$$

for a uniquely determined measure μ on $0 \leqslant \theta \leqslant 1$. In the extreme case where $n = k$, (4.2) gives

$$\Pr\left(\underset{j \leqslant n}{\&} Pa_j\right) = \int_0^1 \theta^n \mu(d\theta). \tag{4.3}$$

Assigning a finite mass μ to the extreme value $\theta = 1$ assures that as n tends to infinity, $\lim_{n \to \infty} \Pr(\&_{j \leqslant n} Pa_j)$ is nonzero. Such an assignment coupled with the axiom of continuity (A4) (chapter 2) assures that $(\forall j)Pa_j$ gets a nonzero prior. Furthermore, since

$$\Pr\left((\forall j)Pa_j \middle/ \underset{j \leqslant n}{\&} Pa_j\right) = \Pr((\forall j)Pa_j)/\Pr\left(\underset{j \leqslant n}{\&} Pa_j\right),$$

we are also assured that a true universal generalization is eventually learned to be true, in that

$$\lim_{n \to \infty} \Pr\left((\forall j)Pa_j \middle/ \underset{j \leqslant n}{\&} Pa_j\right) = 1.$$

Generalizations of de Finetti's representation theorem can be proved for cases where (R1') is relaxed, and the confirmation of universal generalizations can also be studied in these cases.[1]

My conclusion is that Carnap's problem of zero priors for universal generalizations is an artifact of his particular language-based approach to confirmation and not a problem for confirmation theory in general.

2   The Problem of Zero Priors: Jeffrey's Version

While acknowledging that nothing in the probability calculus itself prevents the assignment of nonzero priors to universal laws, Richard Jeffrey nevertheless opines that "in the absence of special reasons to the contrary, it is to be supposed that the [Bayesian] agent's degree of belief in a universal generalization will be zero; for willingness to attribute positive probability to a universal generalization is tantamount to willingness to learn from experience at so great a rate as to tempt one to speak of 'jumping to conclusions'" (1983b, p. 194). This temptation is one that should be resisted, especially by a Bayesian of the personalist school, whose motto is, Have the courage of your convictions. But let us see why Jeffrey thinks that the temptation is there.

By the continuity axiom,

$$\Pr((\forall j)Pa_j) = \lim_{n \to \infty} \Pr\left(\mathop{\&}_{j \leqslant n} Pa_j\right)$$

$$= \lim_{n \to \infty} \Pr(Pa_1) \times \Pr(Pa_2/Pa_1) \times \cdots$$

$$\times \Pr(Pa_n/Pa_1 \ \& \ Pa_2 \ \& \dots \& \ Pa_{n-1}). \tag{4.4}$$

Thus, if the universal generalization is to get a nonzero prior, the factors $\Pr(Pa_n/Pa_1 \ \& \ Pa_2 \ \& \dots \& \ Pa_{n-1})$ must approach 1 sufficiently rapidly as n increases, for otherwise the product of the factors on the right hand side of (4.4) would tend to 0. Just how rapid must this increase be? As an example, Jeffrey sets

$$\Pr(Pa_n/Pa_1 \ \& \ Pa_2 \ \& \dots \& \ Pa_{n-1}) = (2^n - 1)/2^n. \tag{4.5}$$

The product $(1/2)(3/4)(7/8)\dots$ is finite, so the Bayesian agent using this measure will regard $(\forall j)Pa_j$ as confirmable by experience. But this agent also assigns a probability greater than .999 to Pa_{10} after observing 9 positive instances, which means that he is willing to risk \$999 on this outcome to gain \$1, which strikes the faint of heart among us as jumping to conclusions.

Jeffrey's example is only an example, and before drawing any morals from it, we need to know which features are peculiar to it and which carry over to other cases where the generalization is assigned a nonzero prior. The limit in (4.4) is positive just in case each of the factors is positive and

$$\Pr(\neg Pa_1) + \Pr(\neg Pa_2/Pa_1) + \Pr(\neg Pa_3/Pa_1 \ \& \ Pa_2) + \cdots$$

has a finite positive value. If we set $x_n \equiv \Pr(Pa_n/Pa_1 \ \& \dots \& \ Pa_{n-1})$, the sum in question has the form $\sum_{n=1}^{\infty}(1 - x_n)$. Applying the Cauchy ratio test, we find that a sufficient condition for this sum to have a finite positive value is that there exist a constant C such that *for sufficiently large values of n*

$$(1 - x_{n+1})/(1 - x_n) \leqslant C < 1. \tag{4.6}$$

In the case of (4.5), (4.6) is fulfilled, since for every n, $(1 - x_{n+1})/(1 - x_n) = 1/2$. But the qualification that there exist an N such that for all $n > N$, $(1 - x_{n+1})/(1 - x_n) \leqslant C < 1$ means that the ratio can exceed any $C < 1$ for as long a finite stretch as you like and consequently that the rate of learning from experience for this specified stretch can be made to fit any desired criterion of caution.

In sum, there is no need to contemplate assigning zero or infinitesimal priors to universal laws, for willingness to assign finite nonzero priors is not tantamount to a willingness to learn from experience at what the fainthearted may regard as an immodest rate.

3   The Problem of Zero Priors: Popper's Versions

In the infamous appendix 7 ("Zero Probability and the Fine Structure of Probability and Content") of *The Logic of Scientific Discovery*, Popper claims that "*in an infinite universe ... the probability of any (non-tautological) universal law will be zero*" (1961, p. 363). In arguing for this claim, Popper uses a variant of the continuity axiom (A4), although for his purposes it would be enough to note that from the other standard axioms it follows, for example, that $\Pr((\forall j)Pa_j) \leqslant \lim_{n \to \infty} \Pr(\&_{j \leqslant n} Pa_j)$. If we further assume probabilistic independence for distinct instances, i.e.,

$$\Pr(Pa_{n_1} \& Pa_{n_2} \& \dots \& Pa_{n_k}) = \Pr(Pa_{n_1}) \times \Pr(Pa_{n_2}) \times \cdots \times \Pr(Pa_{n_k}) \tag{I}$$

for distinct a's, and if we also impose exchangeability (R1'), then $\lim_{n \to \infty} \Pr(\&_{j \leqslant n} Pa_j) = \lim_{n \to \infty} [\Pr(Pa_k)]^n = 0$ unless $\Pr(Pa_k) = 1$ for any a_k, a condition which not even the most enthusiastic inductivist wants to affirm.

In support of (I) Popper contends that "Every other assumption [opposed to (I)] would amount to postulating *ad hoc* a kind of after-effect; or in other words, to postulating something like a causal connection between [distinct instances]" (1961, p. 367). By using Popper's own notion of propensity probability, we can see, I think, (a) why this contention is false, (b) why Popper's form of anti-inductivism corresponds not to inductive skepticism but to dogmatism, (c) why $\Pr((\forall j)Pa_j) = 0$ is not sufficient by itself for anti-inductivism, and (d) how Hume's ontological scruples can be partially reconciled with one form of inductivism.

Suppose that we know that the repeated trials of an experiment are governed by a fixed propensity p (à la Popper) whose numerical value is as yet undetermined, and that the trials are independent, i.e., IID trials. It is crucial to note that the first 'I' in IID is *not* to be interpreted as independence *tout court*, i.e., as Popper's (I), but rather as a kind of conditional independence. To see what I mean by the latter, assume that the agent who wishes to predict outcomes of the experiment has a probability function $\Pr(\cdot)$ defined on a set $\mathscr{A}$ of propositions that includes propositions about the outcomes and also propositions of the form $p_1 \leqslant p \leqslant p_2$, and assume in addition that the agent associates with every value of $p \in [0, 1]$ a probability function $\Pr_p(\cdot)$ defined on $\mathscr{A}$. Intuitively, $\Pr_p(\cdot)$ is the probability function the agent would adopt if she knew that the value of the propensity is p. $\Pr(\cdot)$ and $\Pr_p(\cdot)$ are knitted together by a consistency condition requiring that for any $A \in \mathscr{A}$,

$$\Pr(A) = \int_0^1 \Pr_p(A)\Pr(dp), \tag{Con}$$

which assumes that $p \mapsto \Pr_p(A)$ is a measurable function for each $A \in \mathscr{A}$.[2] There is nothing new here; in effect, I am just making explicit the assumptions underlying Thomas Bayes's own calculations (see chapter 1). Note that (Con) implies a form of countable additivity. Consider any countable partition of $[0, 1]$ into subintervals $\{I_1, I_2, \ldots\}$. Since $\Pr_{p'}(p \in I_n)$ is 1 or 0 according as p' lies inside I_n or not,

$$\Pr(p \in [0, 1]) = 1 = \int_0^1 \Pr_{p'}(p \in I_n)\Pr(dp') = \sum_{n=1}^{\infty} \Pr(p \in I_n).$$

The 'I' in IID can now be taken to mean that for any k,

$$\Pr_p(Pa_{n_1} \,\&\, Pa_{n_2} \,\&\, \ldots \,\&\, Pa_{n_k}) = \Pr_p(Pa_{n_1}) \times \Pr_p(Pa_{n_2}) \times \cdots \times \Pr_p(Pa_{n_k}). \tag{I$'$}$$

The 'ID' means that $\Pr_p(Pa_j) = p$ for any j. Now try to suppose that Popper's (I) held for this chance setup. In particular, it would mean that

$$\Pr(Pa_{n_1} \,\&\, \ldots \,\&\, Pa_{n_k}/Pa_{n'_1} \,\&\, \ldots \,\&\, Pa_{n'_m}) = \Pr(Pa_{n_1} \,\&\, \ldots \,\&\, Pa_{n_k}).$$

Expanding each side of this equality, we get

$$\frac{\int_0^1 \Pr_p(Pa_{n_1} \& \ldots \& Pa_{n_k} \& Pa_{n_1'} \& \ldots \& Pa_{n_m'})\Pr(\mathrm{d}p)}{\int_0^1 \Pr_p(Pa_{n_1'} \& \ldots \& Pa_{n_m'})\Pr(\mathrm{d}p)}$$

$$= \int_0^1 \Pr_p(Pa_{n_1} \& \ldots \& Pa_{n_k})\Pr(\mathrm{d}p). \tag{4.7}$$

By the IID assumption, (4.7) leads to

$$\int_0^1 p^{k+m}\Pr(\mathrm{d}p) = \int_0^1 p^k\Pr(\mathrm{d}p) \times \int_0^1 p^m\Pr(\mathrm{d}p), \tag{4.8}$$

which can hold for all k and m if and only if the prior probability distribution $\Pr(\mathrm{d}p)$ is concentrated at a point. Thus, imposing (I) on the IID
propensity model requires the agent to be certain from the start about the
value of p, which is surely dogmatism rather than skepticism.

From the point of view of a scientist who has pinpointed the hidden
springs of nature, in this case the value of p, there is no induction, in the
sense that past trials of the experiment do not affect her expectations of
future outcomes. But for us mortals, for whom the hidden springs remain
at least partially hidden, induction takes place not because of some causal
glue that binds together events (there is none) and not because we are
dogmatic rather than skeptical (our skepticism in the sense of open-mindedness may be expressed by a $\Pr(\mathrm{d}p)$ whose support takes in the entire
interval $[0, 1]$). Rather, induction takes place for us because past outcomes
affect our assessment of the unknown p value, and this in turn affects our
assessment of future outcomes. Since the IID assumption and the resulting
(I') provide a precise sense in which the events are causally independent,
and since Pr here is interpreted as personal degree of belief, there is nothing
in the resulting inductivism that should make Hume's hackles rise.

Everything said in the preceding paragraph could be taken as a gloss of
Bayes's model (see chapter 1). Moreover, Bayes's choice of $\Pr(\mathrm{d}p)$ illustrates
how $\Pr((\forall j)Pa_j) = 0$ is not sufficient to rule out inductivism in at least a
weak form. For Bayes's rule implies that $\Pr(\&_{j \leqslant n} Pa_j) = 1/(n + 1)$, which
in turn implies both that $\Pr(Pa_{n+1}/\&_{j \leqslant n} Pa_j) \to 1$ as $n \to \infty$ (successful
instance induction) and that $\Pr((\forall j)Pa_j) = 0$ (zero prior for the general
hypothesis).[3] If only Popper had read Bayes's essay sympathetically, his
philosophical development might have been quite different.

In appendix 7 Popper vacillates between the charge that inductivism
cannot be justified by Hume's lights and the more serious charge that the
inductivism grinds to a halt because of internal inconsistencies in the

mechanism. The former charge is less than telling, since inductivists may feel comfortable in providing a justification that does not fit Hume's strictures or they might opt for a nonjustificatory vindication or they might try to motivate their procedures as a codification of accepted inductive practice in science and declare that any further justification or vindication is unnecessary. The latter and potentially more damning charge is made most clearly in Popper's attempt to give a proof that $\Pr((\forall j)Pa_j)$ must be 0 by hoisting the inductivists on their own petard. Harold Jeffreys (1973) has shown, without using exchangeability (R1') or continuity (A4), that $\Pr((\forall j)Pa_j) > 0$ is a sufficient condition for weak instance induction in the form $\Pr(Pa_{n+1}/\&_{j \leqslant n} Pa_j) \to 1$ as $n \to \infty$ (see section 7 below). Popper claims, in effect, that by applying Jeffreys's result simultaneously to an ordinary predicate 'P' and to a Goodmanized predicate 'P^*', a contradiction is generated. At most, however, what follows is that not both $\Pr((\forall j)Pa_j) > 0$ and $\Pr((\forall j)P^*a_j) > 0$ can hold, not that all universal laws must have zero priors. And on closer inspection it turns out that not even this weaker conclusion follows. This matter will be taken up in detail in section 7 below.

4   The Popper and Miller Challenge

In 1983 *Nature* published a letter by Karl Popper and David Miller containing a purported proof of the impossibility of inductive probability. Although Popper and Miller did not explicitly mention Popper's earlier contention that the prior probability of a universal hypothesis should be zero in a universe with an infinite number of individuals, from a pedagogical point of view their argument is best seen as providing a fallback position for the anti-inductivist. Thus, suppose contra the earlier Popper that the probability of the universal generalization H on the background knowledge K is nonzero. And suppose that evidence E is acquired that raises the probability of H, that is, $\Pr(H/E \& K) > \Pr(H/K)$, which as we know from chapter 3 will be the case when

$$\{H, K\} \models E \quad \text{and} \quad 0 < \Pr(H/K), \Pr(E/K) < 1.$$

Still, Popper and Miller claim, this increase in probability cannot be regarded as confirmation, in the sense of genuine inductive support. To show us why this is so, they invite us to note that for any H and E, H is logically equivalent to $(H \vee E) \& (H \vee \neg E)$. Since E deductively implies the first conjunct, the question of inductive support, they claim, boils down to the way in which E affects the second conjunct. That question is settled by the following two lemmas.

Lemma 1 $\Pr(\neg H/E \mathbin{\&} K) \times \Pr(\neg E/K) = \Pr(H \vee \neg E/K) - \Pr(H \vee \neg E/E \mathbin{\&} K)$ (Popper and Miller)

Proof

$$\Pr(\neg H/E \mathbin{\&} K) \times \Pr(\neg E/K) = (1 - \Pr(H/E \mathbin{\&} K))(1 - \Pr(E/K))$$

$$= 1 - \Pr(E/K) - \Pr(H/E \mathbin{\&} K)$$

$$+ \Pr(H/E \mathbin{\&} K) \times \Pr(E/K)$$

$$= [1 - (\Pr(E/K) - \Pr(H \mathbin{\&} E/K))]$$

$$- \{\Pr(H/E \mathbin{\&} K)\}.$$

The [ ] term can be massaged using a form of total probability: $\Pr(E/K) = \Pr(H \mathbin{\&} E/K) + \Pr(\neg H \mathbin{\&} E/K)$. Thus $\Pr(E/K) - \Pr(H \mathbin{\&} E/K) = \Pr(\neg(H \vee \neg E)/K) = 1 - \Pr(H \vee \neg E)$, which shows that the [ ] term is equal to $\Pr(H \vee \neg E/K)$. Since the { } term is equal to $\Pr(H \vee \neg E/E \mathbin{\&} K)$, the lemma is proved.

As a direct consequence of Lemma 1 we get lemma 2:

Lemma 2   If $\Pr(H/E \mathbin{\&} K) \neq 1 \neq \Pr(E/K)$, then $\Pr(H \vee \neg E/E \mathbin{\&} K) < \Pr(H \vee \neg E/K)$.

According to Popper and Miller, lemma 2 is "completely devastating to the inductive interpretation of the calculus of probability" (1983, p. 688). If $\Pr(E/K) = 1$, $\Pr(H/E \mathbin{\&} K) = \Pr(H/K)$, and E does not incrementally support H. The case where $\Pr(H/E \mathbin{\&} K) = 1$ is also uninteresting, for if Pr is strictly coherent (see chapter 2), $E \mathbin{\&} K \models H$, which places us in the realm of deductive rather than inductive support. And even if strict coherence is rejected, the case is still uninteresting, since $\Pr(H/E \mathbin{\&} K) < 1$ for most real-life examples of confirmation in science. Thus, we may safely assume that the conditions of lemma 2 are satisfied, and we may conclude that $H \vee \neg E$, what Popper and Miller call the part of H that goes beyond E, is countersupported by E.

Popper and Miller introduce their letter with an unfortunate flourish: "Proofs of the impossibility of induction have been falling 'dead-born from the Press' since the first of them (in David Hume's *Treatise of Human*

Nature) appeared in 1739. One of us (K. P.) has been producing them for more than 50 years. This one strikes us both as pretty" (1983, p. 687). Hume did not claim to have proved the impossibility of induction but rather that induction cannot be justified except on a psychological basis. By contrast, Popper and Miller offer an impossibility proof that does literally fall dead-born from the press. For once the assumption of nonzero priors is granted, there is no firm ground for the anti-inductivist to stand on, or so I will argue.

The Popper and Miller article has elicited a sizable number of comments, both pro and con, but many of the relevant points can be discerned from the first responses published in *Nature* by Isaac Levi (1984) and Richard Jeffrey (1984). Levi endorsed the Popper and Miller position by appeal to the principle that if $E \& K \models H_1 \leftrightarrow H_2$, then $\Pr(H_1/E \& K) = \Pr(H_2/E \& K)$. Applying this to the factorization of H into $(H \vee E) \&$ $(H \vee \neg E)$, we note that $E \models H \leftrightarrow (H \vee \neg E)$, so that we should have $\Pr(H/E \& K) = \Pr(H \vee \neg E/E \& K)$. This is perfectly correct. But as Gaifman (1985) warns, the absolute concept of confirmation must not be confused with the incremental one (see chapter 3). From $E \models H_1 \leftrightarrow H_2$ it does not follow that E has the same incremental effect on both hypotheses, i.e., that $\Pr(H_1/E \& K) - \Pr(H_1/K) = \Pr(H_2/E \& K) - \Pr(H_2/K)$, even if K is a tautology. Indeed, the Popper and Miller lemmas provide a counterexample, since they show that $\Pr(H \vee \neg E/E \& K) < \Pr(H \vee \neg E/K)$, even in the case where $\Pr(H/E \& K) > \Pr(H/K)$.

Jeffrey rejected the Popper and Miller argument on the grounds that it rests on a specious identification of $H \vee \neg E$ as the part of H that goes beyond the evidence E. To take a concrete example, the part of $(\forall j)Pa_j$ that goes beyond $Pa_1 \& Pa_2 \& Pa_3$ is intuitively $(\forall j)[(j \geqslant 4) \rightarrow Pa_j]$ and not the Popper and Miller $(\forall j)Pa_j \vee \neg(Pa_1 \& Pa_2 \& Pa_3)$.

Gillies (1986) responded that the Popper and Miller argument can be restated in a form that is independent of the identification of $H \vee \neg E$ as the part of H that goes beyond E. Suppressing K for the sake of simplicity, define the support given by E to H as $s(H, E) \equiv \Pr(H/E) - \Pr(H)$. It is then easy to show that $s(H, E) = s(H \vee E, E) + s(H \vee \neg E, E)$. Since, according to Gillies, $s(H \vee E, E)$ represents the deductive support conferred by E, $s(H \vee \neg E, E)$ must represent the inductive support, which we have already seen to be negative. So far so good, but what is needed to complete the impossibility proof is an inference that moves from the facts that the support function $s(H, E)$ is the sum of two functions and that neither of the latter functions can represent positive inductive support to the conclusion that $s(H, E)$ cannot represent positive inductive support. As Chihara (1988) notes, the following general inference rule (G) is not valid:

G   If function f is thought to represent F and if $f = f_1 + f_2$, where neither f_1 nor f_2 can represent F, then f cannot represent F.

Thus the burden is on the critic of inductivism to show that the relevant instance of (G) is truth-preserving.[4]

Aside from the details of the debate between Popper-Miller and their critics, the real question for inductivism is the one emphasized by Nelson Goodman in *Fact, Fiction, and Forecast*, namely, when do already observed instances confirm a hypothesis merely by content cutting (i.e., by entailing part of the content of the hypothesis) and when do they genuinely confirm it in the sense of supporting its predictions about unexamined instances?[5] To make the question concrete, consider again $H: (\forall j)Pa_j$, and let Pa_1 & Pa_2 & ... & Pa_n stand for the observed instances. We want to know whether the probability for a positive result for the next instance Pa_{n+1} or the next m instances $Pa_{n+1}, Pa_{n+2}, ..., Pa_{n+m}$ increases as n increases and, if so, whether the increase continues until certainty is reached in the limit. Popper's original form of anti-inductivism attempted to derive negative answers from claims about zero priors. By contrast, the Popper and Miller argument attempts to prove the impossibility of inductivism even granting the inductivist that $(\forall j)Pa_j$ has a positive prior. But that concession makes the answers to our questions *demonstrably yes*, even if we assume nothing about exchangeability or continuity (see section 7 below).

5   Richard Miller and the Return to Adhocness

As noted in chapter 3, Quine's holism and his thesis that any scientific theory can be rationally maintained come what may are derived by combining a bit of Duhem with a lot of HD methodology and a refusal to entertain any other constraints on theory testing. Duhem's point was that an HD test of a typical theory T requires the use of auxiliary assumptions A consisting, perhaps, of other theories plus boundary and/or initial conditions, for it is not T by itself but T plus A that entails a prediction E that can be submitted to the judgment of observation and experiment. In the face of a negative judgment the theory T can be saved by blaming the outcome $\neg E$ on errors in A. If deductive logic alone were the only tool available for assessing the bearing of evidence on theories, it would indeed seem to follow that T can be maintained, come what may.

In chapter 3 Bayesianism was touted as a potential antidote to Duhem and Quine methodological pessimism, though there was some hand waving about the problem of objectivity, whose discussion is postponed to chapter 6. Richard Miller (1987) has argued that even if this hand waving is permitted to pass, Bayesianism is not the panacea it might seem. Indeed, he thinks that the HD problem of adhocing the auxiliaries revisits Bayesianism as the problem of adhocing the likelihoods. Miller puts his complaint in the form of a dilemma: "If the likelihoods were kept rigid, rational belief in many good theories would have to be lowered to an implausible degree. If likelihoods can be bent to fit the facts, Bayes' Theorem cannot account for the real force of compelling arguments" (1987, p. 319).

Let us try to understand each horn of Miller's dilemma. To begin with the second horn, it is not Bayes's theorem itself but the combination of the theorem and a rule of conditionalization (chapter 2) that accounts for the force of compelling arguments. When E is learned, the combination gives, in the case of strict conditionalization,

$$\mathrm{Pr}_{\mathrm{new}}(T) = \mathrm{Pr}_{\mathrm{old}}(T/E) = (\mathrm{Pr}_{\mathrm{old}}(T) \times \mathrm{Pr}_{\mathrm{old}}(E/T))/\mathrm{Pr}_{\mathrm{old}}(E).$$

Thus, if $\mathrm{Pr}_{\mathrm{old}}(E/T) \ll \mathrm{Pr}_{\mathrm{old}}(E)$, learning E leads to the result that $\mathrm{Pr}_{\mathrm{new}}(T) \ll \mathrm{Pr}_{\mathrm{old}}(T)$. This much is clear. What Miller means by "bending" or "revising" the likelihoods is less clear, but his discussion of the use of auxiliary hypotheses suggests the following interpretation. Suppose that $\{A_i\}$ is a partition of the relevant auxiliaries that could be brought into play. Then by the principle of total probability,

$$\mathrm{Pr}_{\mathrm{old}}(E/T) = \sum_i \mathrm{Pr}_{\mathrm{old}}(E/T \,\&\, A_i) \times \mathrm{Pr}(A_i/T)$$

and

$$\mathrm{Pr}_{\mathrm{old}}(E) = \sum_i \mathrm{Pr}_{\mathrm{old}}(E/A_i) \times \mathrm{Pr}_{\mathrm{old}}(A_i).$$

Insofar as the factors $\mathrm{Pr}_{\mathrm{old}}(E/A_i \,\&\, T)$ are "objective"—as they are, for instance, when $T \,\&\, A_i \models E$ or $T \,\&\, A_i \models \neg E$—a change of heart about the likelihood of E on T will have to come through a change of heart about $\mathrm{Pr}_{\mathrm{old}}(A_i/T)$.[6] Fiddling these factors along with the factors $\mathrm{Pr}_{\mathrm{old}}(A_i)$ can ameliorate or cancel the original judgment that $\Pr_{\text{old}}(E/T) \ll \Pr_{\text{old}}(E)$ and thus that $\Pr_{\text{new}}(T) \ll \Pr_{\text{old}}(T)$.

While such fiddlings might be termed Bayesian adhocery, they are more accurately described as an abandonment of Bayesianism. In chapters 5 and 8, I will argue that data-instigated hypotheses call for literally ad hoc responses. When new evidence suggests new theories, a non-Bayesian shift in the belief function may take place. But leaving such cases aside for the moment, any Bayesian worth his salt will want some compelling reason for abandoning his credo. Miller's reason is that there are situations where either likelihoods are bent or else rational belief in a good theory would have to be lowered to an implausible degree. This claim is itself subject to a dilemma.

Suppose first that a "good theory" is one deserving of attention because it has some desirable features, such as fecundity, predictive power, simplicity, or whatnot. Then there is no presumption that such good theories always or ever have significant posterior probabilities. Indeed, if the competition is keen and there are many incompatible good theories vying for our allegiance, then the probability calculus alone assures that some of them must have small probabilities. Suppose next that a "good theory" is one that, on the basis of the total available evidence, has a high posterior probability either in absolute terms or relatively to competing theories. Then it is analytic that belief in good theories cannot be lowered to an implausible degree.

In effect, Miller's contention is that the second horn of this second dilemma is unsound because the history of science provides examples where there is a clash between Bayesian calculations on the one hand and intuitions about what is to be believed on the other. In response, the Bayesian can offer two diagnoses. First, insofar as the clash is real, it must be resolved contra intuitions. In the Bayesian game you are allowed to play favorites once by your assignments of prior probabilities. Let us suppose that you did so in assigning T a high prior (because of its many making-good features) and also that you agreed that E is more likely by itself than on the assumption of T. But now upon learning that E, you want to renege on the latter commitment so that you can continue to assign a high probability to T. The Bayesian machinery prevents this second round of playing favorites, *and so it should*. Second, the Bayesian can make the case that the clash that Miller tries to promote does not occur in actual cases. Given our present evidence about geology and population genetics, we *now* take Darwin's theory of evolution to be a good theory in the worthy-of-belief sense. But there is no presumption that theories now accorded this status must always have had it at every earlier stage of inquiry. Indeed, the gaps in the fossil record, the existence of fossils in the lower sedimentary levels, and many other difficulties that Darwin himself pointed out should have meant that in Darwin's day his theory had a low posterior. The story of how the accumulation of additional evidence raised the posterior probability of the theory of evolution is a paradigm example of progress in science. The adhocery Miller seems to advocate would disguise this progress.

Science, like our system of justice, assumes that the truth is most likely to out if adversaries prosecute the case and a decision is rendered by an impartial judiciary. A crucial difference is that in the scientific enterprise the same scientists play both roles at once. This perhaps encourages Miller, as it encourages scientists themselves, to blur the difference between the two roles. Scientists qua advocates of competing theories are free to blame the difficulties in their pet theories on the auxiliaries and to bend the likelihoods if they think this will help their advocacy. But the reaction of scientists qua neutral judges should be not to adopt non-Bayesian attitudes but to change probabilities in a Bayesian fashion in the light of whatever further evidence the advocates can provide. Admittedly, the two roles are hard to keep separate, but not to strive for a separation is a prescription not just for anti-Bayesianism but also for antiscience and antirationality.

None of the above touches a more traditional form of adhocery. If T gets into genuine Bayesian trouble by having its posterior probability driven further and further downward by accumulating evidence, then T *itself* can be adhoced, i.e., T can be modified in those features responsible for the diminishing probability. But there should be nothing in Bayesianism to prevent such a tactic; indeed, not only is it not a dishonorable tactic to modify theories in the light of adverse evidence, it is good scientific common sense. Of course, if successive adhocings continue to run into trouble, the program of trying to keep alive the theory takes on a degenerative cast. When to abandon what seems to be a sinking ship and when to try to keep the ship afloat is a difficult and delicate problem.

A related problem is suggested by Miller's Darwin example, although the problem itself is hardly peculiar to biology.[7] When Darwin first sprang "the theory of evolution" on the world, what he was offering was not a concrete theory but a general type of theory. Initially neither he nor his supporters could supply a specific version of the theory that deserved to be accorded a high posterior probability on the basis of the available evidence. But clearly, the theory type the Darwinians were proposing was worth pursuing—or so it seems in retrospect. The general problem of when a theory type, all of whose extant versions have a low or even flatly zero posterior, should be pursued and when it should be cast on the trash heap of science is, like its cousin mentioned above, a difficult one. Bayesianism offers no solutions to such problems, nor should it. The problems are ones of practical decision. What Bayesianism can offer is an account of one of the essential ingredients that go into the decision-making process. In the Darwinian case a key ingredient is the probability of the vague hypothesis that some theory of evolution that is recognizably Darwinian (say, because it assigns natural selection a central role in descent) is true or approximately true. The other essential ingredient is the collection of utilities assigned to the alternative courses of action. It is then a matter of choosing the action that maximizes the expected utility. Beyond this, I very much doubt that there are any valid and nonplatitudinous methodological rules to be stated about such decision problems. I will return to these matters in chapter 8.

6 Grünbaum's Worries

Popper worried that the Bayesian machinery does not work at all. Grünbaum (1976) worried that it works too well. His worries are encapsulated in a series of challenges. "Can the Bayesian succeed where unbridled instantianist inductivism failed, and show why the same confirming experiment should *not* be repeated *ad nauseam*?" (1976, p. 242). Grünbaum's demand is that the Bayesians meet this challenge by supplying a proof of "a *monotonic* diminution in the amounts of probability increase" from repetitions of the same experiment (1976, p. 242). But if we rule out the case where convergence to certainty takes place after a finite number of experimental results have come in, it follows trivially that the successive amounts of probability increase must eventually diminish, and this is so whether the results come from repetitions of the same experiment or from performances of different experiments. Of course, one's intuition is that the diminution ought to set in sooner and ought to be steeper in the former cases than in the latter cases. Instances of this intuition cannot be demonstrated from the probability calculus alone, and in that sense Bayesians cannot meet Grünbaum's challenge. But if judgments as to variety of evidence are not a priori but depend upon judgments about how the world is organized, then the Bayesian should not be embarrassed in rejecting the stringent form of Grünbaum's challenge. There would most certainly be an embarrassment here if the Bayesian apparatus could not be used to analyze variety of evidence, but as I tried to indicate in chapter 3, the apparatus appears to be promising in this regard.

Grünbaum also issued a second related challenge:

Does Bayesianism sanction the credibilification of a new hypothesis by a sufficient number of positive instances which are the results of NON-risky predictions? For example, would a sufficient number of cases of people afflicted by colds who drink coffee daily for, say, two weeks and recover not confer a *posterior* probability greater than 1/2 on the new hypothesis that such drinking *cures* colds? (1976, p. 243)

He is pessimistic that the Bayesian can respond in the negative, for apart from some restricted cases, he writes, the Bayesian formalism will typically permit the following: "In the case of those successful but *non*-risky predictions which *are* supportive, a sufficient number of them will credibilify a hypothesis to be more likely true than not" (1976, p. 244). The qualifier "non-risky" is crucial, since without it the claim is demonstrably false for theoretical hypotheses or more generally for the case where many competing hypotheses cover the same data. This is easily shown in the HD case where the predictions are deductively entailed by the hypothesis and the background information.

Counterclaim 1 Suppose that $\{H, K\} \models E_i$, $i = 1, 2, \ldots.$ Then it is *not* the case that $\Pr(H/E_1 \& E_2 \& \ldots \& E_n \& K) \to 1$ as $n \to \infty$ if there is an alternative H' such that $\{H', K\} \models E_i$ for each i, $K \models \neg(H \& H')$, and $\Pr(H'/K) > 0$.

Proof By Bayes's theorem,

$$\frac{\Pr(H/E_1 \& E_2 \& \ldots \& E_n \& K)}{\Pr(H'/E_1 \& E_2 \& \ldots \& E_n \& K)} = \frac{\Pr(H/K)}{\Pr(H'/K)}.$$

Since H and H' are incompatible, given K, $\Pr(H/X \& K) + \Pr(H'/X \& K) \leqslant 1$. So if the posterior probability of H were to go to 1, the posterior of H' would go to 0, with the result that $(\Pr(H/K)/\Pr(H'/K)) = +\infty$ and thus that $\Pr(H'/K) = 0$.

Counterclaim 2   Suppose that $\{H, K\} \models E_i, i = 1, 2, \ldots$ Then $\Pr(H/E_1 \,\&\, E_2 \,\&\, \ldots \,\&\, E_n \,\&\, K) < .5$ for all n if there is a rival H' such that $\{H', K\} \models E_i$ for each i, $K \models \neg(H \,\&\, H')$, and $\Pr(H'/K) > \Pr(H/K)$.

Proof   Similar.

Perhaps Grünbaum's claim will hold when the qualification 'nonrisky' is restored. But on behalf of the Bayesian, I ask for the precise sense of 'nonrisky' to be supplied, and I predict that when this is done, either the claim will not prove to be correct, or else the resulting Bayesian credibilification will not be counterintuitive.

In fairness to Grünbaum, I should note that I have not responded to the core of his worry that Bayesian methods are not helpful in evaluating causal hypotheses, such as that drinking coffee cures colds. "And how, if at all," he asks, "does the Bayesian conception of inductive support enjoin scientists to employ a *control group* in the case of this causal hypothesis ... with a view to testing the rival hypothesis that coffee consumption is causally irrelevant to the remission of colds?" (1976, p. 243). I am of the conviction that causal talk is a mare's nest of confusions, snares, and delusions. In any case, I insist that a prelude to successful hypothesis testing is a precise statement of the hypothesis in noncausal language. Thus, perhaps what is meant by the claim that coffee consumption cures colds is simply that the rate of remission among people who drink coffee is higher than among those who don't, other factors being constant. If so, the need for a control group is self-evident. Or perhaps the claim is meant to point to a hypothesis about the action of some chemical ingredient of coffee either on the immune system or directly on cold viruses. But again, this version of the hypothesis needs to be stated precisely either as a universal or a statistical generalization, without weasel causal words. And once this is done, I am confident that Bayesianism is competent to illuminate the methodology of testing the generalization.

7   Goodman's New Problem of Induction

Enough ink has been spilled over Goodman's "new problem of induction" to drown an elephant. Not surprisingly, most of the spill is directed at Goodman's notion of "entrenchment" and his attempts to use this notion to fashion rules of projectability. What is more than a little surprising is the paucity of literature devoted to clarifying what the new problem is supposed to be. The need for such a clarification will, I trust, soon become evident.

Goodman takes Hume to task not for his descriptive approach to the problem of induction but for "the imprecision of his description":

Regularities in experience, according to [Hume], give rise to habits of expectation.... But Hume overlooks the fact that some regularities do and some do not establish such habits.... Regularities are where you find them, and you can find them anywhere.... Hume's failure to recognize and deal with this problem has been shared even by his most recent successors. (1983, p. 82)

According to Goodman, the remaining problem left unresolved by Hume is to describe which hypotheses are and which are not capable of receiving confirmation from their instances.

To begin the discussion of this form of Goodman's problem, suppose that instances are taken in the HD sense, i.e., the instances E_i, $i = 1, 2, \ldots$, are deductive consequences of hypothesis H and the background knowledge K. (If you want H to be a universal conditional, such as $(\forall i)(Pa_i \to Qa_i)$, take the instances E_i to be of the form $(Pa_i \to Qa_i)$, or else let K state that all the objects examined are P's and take the instances to be of the form $Pa_i \ \& \ Qa_i$.) Then we know that H is confirmable by its instances in the sense that $\Pr(H/E_i \ \& \ K) > \Pr(H/K)$, at least if

$$0 < \Pr(H/K), \Pr(E_i/K) < 1.$$

So if we exclude the cases where H or E_i is already known to be almost surely true or almost surely false, Goodman's new problem would seem to boil down to identifying the hypotheses to which nonzero priors are to be attached. Unfortunately, this formulation does not accord with the examples Goodman gives and in particular with his claims that the hypothesis "All men now in this room are third sons" is not confirmed by the finding that some particular man now in this room is a third son, and that "All emeralds are grue" is not confirmed by the evidence that emeralds examined before the year 2000 are all green.[8] For it is much too draconian to suppose that these and other hypotheses in the target class of unprojectable hypotheses are to be initially and forever condemned to limbo by receiving zero priors. (If this is right, it follows that the difference between a projectable hypothesis and a nonprojectable one cannot lie merely in the difference in the prior probabilities of these hypotheses. Of course, if belief change goes via conditionalization, with $\Pr_{new}(\cdot) = \Pr_{old}(\cdot/E)$, the difference must lie in the prior probabilities in the sense of the entire $\Pr_{old}(\cdot)$

distribution. But it remains to pinpoint more precisely the features of $\mathrm{Pr}_{\mathrm{old}}(\cdot)$ that make for the difference. That is the task undertaken below.)

A natural response to such examples is to distinguish, as suggested above in section 4, between genuine confirmation and mere content cutting. The third son and the grue hypotheses may be confirmed, in the sense of having their probabilities raised, by the positive instances in question. But intuitively, the probability goes up only because part of the content of the hypotheses has been exhausted, and the past instances do not serve to boost confidence that future instances will conform to the predictions of the hypothesis. Goodman himself puts the point this way: genuine confirmation occurs "only when an instance imparts to the hypothesis some credibility that is conveyed to other instances" (p. 69). In keeping with Goodman's terminology, let me apply the term 'projectable' to a hypothesis capable of receiving such genuine confirmation from its instances.[9]

Bayesians will not be easily convinced that they have been presented with a persuasive distinction, for the new understanding of the new problem seems to leave us with the same conundrum as before. Let us say that relative to K, H is *weakly* (respectively, *strongly*) *projectable in the future-moving sense* over its instances $E_1, E_2, \ldots$ just in case

$$\lim_{n \to \infty} \mathrm{Pr}\left(E_{n+1} \,\middle/\, \underset{i \leqslant n}{\&}\, E_i \,\&\, K \right) = 1$$

$$\left(\text{respectively,} \; \lim_{m,n \to \infty} \mathrm{Pr}\left(\underset{n < j \leqslant n+m}{\&}\, E_j \,\middle/\, \underset{i \leqslant n}{\&}\, E_i \,\&\, K \right) = 1 \right).[10]$$

Similarly, we can say that relative to K, the predicate 'P' is *weakly* (respectively, *strongly*) *projectable in the future-moving sense* over the individuals $a_1, a_2, \ldots$ just in case

$$\lim_{n \to \infty} \mathrm{Pr}\left(Pa_{n+1} \,\middle/\, \underset{i \leqslant n}{\&}\, Pa_i \,\&\, K \right) = 1$$

$$\left(\text{respectively,} \; \lim_{m,n \to \infty} \mathrm{Pr}\left(\underset{n < j \leqslant n+m}{\&}\, Pa_j \,\middle/\, \underset{i \leqslant n}{\&}\, Pa_i \,\&\, K \right) = 1 \right).$$

Now a sufficient condition for both the weak and strong projectability of H is that $\mathrm{Pr}(H/K) > 0$.

Proof of the sufficiency for weak projectability (H. Jeffreys) By Bayes's theorem and the fact that $\{H, K\} \models E_i$,

$$\Pr\left(H \middle/ \underset{i\leqslant n+1}{\&} E_i \& K\right)$$

$$= \frac{\Pr(H/K)}{\Pr(E_1/K) \times \Pr(E_2/E_1 \& K) \times \cdots \times \Pr(E_{n+1}/\underset{i\leqslant n}{\&} E_i \& K)}.$$

If $\Pr(H/K) > 0$, the denominator on the right-hand side will eventually become smaller than the numerator, which contradicts an axiom of probability, unless $\Pr(E_{n+1}/\underset{i\leqslant n}{\&} E_i \& K) \to 1$ as $n \to \infty$.

Proof of the sufficiency for strong projectability (Huzurbazar) Rearrange Bayes's theorem to read

$$\Pr\left(\underset{i\leqslant n}{\&} E_i/K\right) = \Pr(H/K)/\Pr\left(H \middle/ \underset{i\leqslant n}{\&} E_i \& K\right).$$

Setting $u_n \equiv \Pr(\underset{i\leqslant n}{\&} E_i/K)$, we see that $u_n \geqslant \Pr(H/K) > 0$. Since $u_{n+1} = u_n \times \Pr(E_{n+1}/\underset{i\leqslant n}{\&} E_i \& K)$ and $u_1, u_2, \ldots$ is a monotone decreasing sequence bounded from below, it must have a limit $L \geqslant \Pr(H/K) > 0$. Thus, $\lim_{m,n\to 0}(u_{n+m}/u_n) = L/L = 1$.

As a result, $\Pr((\forall i)Pa_i/K) > 0$ is a sufficient condition for both weak and strong projectability of 'P'.

We know from previous sections that $\Pr((\forall i)Pa_i/K) > 0$ is not a necessary condition for weak projectability of 'P'. We can say a bit more about this matter under the assumption of exchangeability (R1′). Applying de Finetti's representation theorem (section 2), we see that the necessary and sufficient condition for the failure of weak projectability is

$$\lim_{n\to\infty} \frac{\int_0^1 \theta^{n+1}\mu(\mathrm{d}\theta)}{\int_0^1 \theta^n\mu(\mathrm{d}\theta)} < 1. \tag{CM}$$

The label (CM) is chosen to indicate a closed-minded attitude, for (CM) is equivalent to the condition that $\mu([0, \theta^*]) = 1$ for some $\theta^* < 1$, which rules out any possibility for an instance of 'P' to have a probability greater than θ^*. The extreme case of closed-mindedness is a μ concentrated on a point. For example, if $\mu(\{1/2\}) = 1$, then each instance of 'P' is assigned a probability of $1/2$ independently of all the other instances. Thus the user of the resulting Pr function is certain of the probability of an instance of 'P', so certain that no number of other instances will change her mind. The probability measure advocated by Wittgenstein in the *Tractatus*, already encountered in chapter 1, had this character.

Another sense of projectability for predicates sometimes used in the literature is codified in the definition that relative to K, 'P' is *somewhat projectable in the future-moving sense* over the individuals $a_1, a_2, \ldots$ just in case for each n,

$$\Pr\left(Pa_{n+1} \Big/ \underset{i \leqslant n}{\&} Pa_i \,\&\, K\right) > \Pr\left(Pa_n \Big/ \underset{i \leqslant n-1}{\&} Pa_i \,\&\, K\right).$$

Under exchangeability, this sense of projectability holds unless μ is concentrated on some value of θ, as can be seen by applying the Cauchy and Schwartz inequality. Thus the case of a closed-minded but not completely closed-minded μ provides an example where 'P' is somewhat but not weakly projectable.

To return from this digression, the main point is that we still have failed to locate a persuasive "new problem" of induction. Projectability as construed above fails to separate green from grue, except on the implausible assumption that the grue hypothesis is to be assigned a flatly zero prior and the green hypothesis a nonzero prior.

Perhaps the failure is due to using the wrong definition of projectability. Let us try a somewhat different tack by introducing a doubly infinite array of individuals: $\ldots, a_{-n}, a_{-n+1}, \ldots, a_0, a_1, \ldots, a_{n-1}, a_n, \ldots$. The axiom of continuity, should we choose to employ it here, would be written as

$$\Pr((\forall i)Pa_i) = \lim_{n \to \infty} \Pr\left(\underset{-n \leqslant i \leqslant n}{\&} Pa_i\right). \tag{A4$'$}$$

And we can say that relative to K, 'P' is *weakly projectable in the past-reaching sense* just in case for any n,

$$\Pr(Pa_{n+1}/Pa_n \,\&\, Pa_{n-1} \,\&\, \ldots \,\&\, Pa_{n-k} \,\&\, K) \to 1$$

as $k \to +\infty$, and similarly for strong projectability. To illustrate the difference between the past-reaching and the future-moving senses of projectability, think of the individuals as a doubly infinite sequence of days stretching infinitely into the past and the future, and take Pa_i to mean that the sun rises on day i. Then the past-reaching sense of projectability requires that one's confidence that the sun will rise on some *fixed* future day approaches certainty as the experience of sunrises stretches further and further into the past, whereas the future-moving sense requires only that one's confidence in a new sunrise approaches certainty for a shifting future day that recedes into the future along with the accumulating instances of new dawns. Practical inductive inference requires that we stand pat in the present and lay our bets on the happenings of a fixed future day. Thus the unresolved component of Hume's problem can arguably be identified with the task of describing the hypotheses and predicates we take to be projectable in the past-reaching sense. We have finally arrived at a real problem. For projectability of a predicate to hold in the future-moving sense, it is sufficient for the universal generalization over the predicate to receive a nonzero prior. But not so for the past-reaching sense of projectability, as Goodman's own examples show. Consider a predicate 'P' and a grue analogue 'P^*' defined by

$$P^*a_i \equiv [((i \leqslant 2000) \mathbin{\&} Pa_i) \vee ((i > 2000) \mathbin{\&} \neg Pa_i)],$$

where i ranges from 1 to $+\infty$. Nothing in the probability calculus prevents us from assigning nonzero priors to both $(\forall i)Pa_i$ and $(\forall i)P^*a_i$, in which case it follows both that $\Pr(Pa_{n+1}/\mathbin{\&}_{i \leqslant n} Pa_i) \to 1$ and that $\Pr(P^*a_{n+1}/\mathbin{\&}_{i \leqslant n} P^*a_i) \to 1$ as $n \to \infty$. But contrary to what Popper suggests in appendix 7 of *The Logic of Scientific Discovery* (see section 2 above), no contradiction results, since the rates of convergence need not be the same. Indeed, Goodman's examples prove that the rates of convergence cannot be uniform over different predicates (see Howson 1973).

Now let i range between $-\infty$ and $+\infty$. The assignment of positive priors to both $(\forall i)Pa_i$ and $(\forall i)P^*a_i$ does not guarantee, on pain of contradiction, that 'P' and 'P^*' are both projectable in the past-reaching sense, for otherwise we would have both

$$\Pr(Pa_{2001}/Pa_{2000} \mathbin{\&} Pa_{1999} \mathbin{\&} \ldots \mathbin{\&} Pa_{2000-k}) \to 1$$

as $k \to +\infty$ and

$$\Pr(P^*a_{2001}/P^*a_{2000} \mathbin{\&} P^*a_{1999} \mathbin{\&} \ldots \mathbin{\&} Pa^*a_{2000-k}) \to 1$$

as $k \to +\infty$, which is a contradiction, since P^*a_{2001} is equivalent to $\neg Pa_{2001}$, while $P^*a_{2000}, P^*a_{1999}, \ldots$ are equivalent to $Pa_{2000}, Pa_{1999}, \ldots$.

Whether the line between projectable and nonprojectable hypotheses in the past-reaching sense is to be drawn in terms of entrenchment or the like is an issue I will not broach here. What I will note is that if exchangeability holds for 'P', past-reaching projectability for 'P' is equivalent to future-moving projectability. This occasions two further remarks. The first is that if we assign nonzero priors to both $(\forall i)Pa_i$ and $(\forall i)P*a_i$, then we cannot consistently take exchangeability to hold for both 'P' and '$P*$'. Conversely, if we take exchangeability to hold for both, then consistency demands that one of the measures μ and $\mu*$ in de Finetti's representation theorem be closed-minded. The second remark is that in the demonstrations of projectability, exchangeability is functioning as a uniformity-of-nature postulate. This allows Goodman to revive his original point: uniformities are where you find them, and you find them everywhere. The Bayesian apparatus allows Goodman's problem to be stated in a more precise form. If we want to be open-minded, then we have to make careful choices of where to see uniformity in the sense of exchangeability.

Another nice feature of the analysis is that Hume's old problem and Goodman's new problem are seen to be facets of the general problem of underdetermination of hypotheses by evidence. In chapter 6, I will consider the extreme form of this problem where theoretical hypotheses are underdetermined by all possible observational evidence. As a result of this underdetermination, no amount of evidence will force a merger of opinion or convergence to certainty about the true hypothesis. Here we are confronted with a more limited but no less interesting form of the problem: the underdetermination of observational hypotheses about the future by all possible evidence about the past. As a result, no amount of past evidence forces convergence to certainty about the true hypothesis, at least not without the help of substantive assumptions, such as exchangeability.

The above remarks help us to appreciate an important but now largely forgotten exchange between Goodman and Carnap in 1946 and 1947. Goodman's "Query on Confirmation" (1946), directed primarily at Hempel and Carnap, concluded that what these authors had given us was "an ingenious and valuable logico-mathematical apparatus that we may apply to the sphere of projectable or confirmable predicates whenever we discover what a projectable or confirmable predicate is" (p. 385). Carnap's response came in a paper entitled "On the Application of Inductive Logic" (1947). There he demanded that the qualities or relations designated by the primitive predicates of his confirmation language be "simple," i.e., not analyzable into simpler components. *Purely qualitative* properties (e.g., 'blue') are then defined as those that can be expressed without the use of individual constants but not without the help of primitive predicates. *Purely positional* properties (e.g., '$x = a_{28}$'), by contrast, can be expressed without the help of primitive predicates. And finally, *mixed properties* (e.g., 'x is red or $x = a_{28}$') are ones that fall into neither of the previous categories. Carnap tentatively proposed that all purely qualitative properties are projectable, while purely positional ones are not, and he was inclined to think mixed properties are not, "but this requires further investigation" (p. 146). Goodman's charge of incompleteness is, then, rebutted, since all the primitive predicates of the confirmation language are simple and therefore projectable, and as for the complex predicates, it is only a matter of applying c^* (which was then Carnap's favorite confirmation function) and seeing what the results are.[11]

The reader familiar with Goodman's style can easily guess the gist of his rejoinder, published as "On the Infirmities of Confirmation-Theory" (1947). Goodman began by rejecting Carnap's root assumption that there are absolutely simple properties. "The nature of this simplicity is obscure to me since the question of whether or not a given property is analyzable seems to me to be quite as ambiguous as the question whether a given body is in motion" (p. 149). Goodman then complained that

we are offered no evidence or argument in support of Carnap's conjecture that either the class of purely qualitative predicates is identical with the class of intuitively projectable predicates, or that such predicates as are projectable though not purely qualitative will also prove to be projectable by his definition. The first alternative seems *prima facie* dubious since predicates like "solar," "arctic," and "Sung" appear to be intuitively projectable but are not purely qualitative; the grounds for the second alternative are not evident. (Pp. 149–150)

We can now see that something is to be said for either side in this debate. In favor of a Carnapian or more generally a probabilistic approach is that the use of the probability calculus, so conspicuously absent in Goodman's writing on induction, leads to a significant clarification of the problem of projectability.[12] Specifically, it allows us to distinguish different senses of projectability and to prove results about the conditions under which projectability in one or more of these senses holds or fails. As a case in point, we have seen that if the "purely qualitative" nature of a predicate 'P' is taken to imply exchangeability (and Carnap so took it) and the μ of de Finetti's representation theorem is not closed-minded, then no Goodmanian excursions into the pragmatics of entrenchment or the like are needed, for it is demonstrable that 'P' is projectable.

Such results, however, do not entirely justify the closing paragraph of Carnap's paper:

If somebody were to criticize an axiom of Euclidean geometry because it does not contain a rule specifying for which particular class of triangles the Pythagorean theorem holds, the author of the system might reply: no additional rule is required; the axiom system is complete; take it and discover for yourself under what conditions the theorem holds. If anybody misses in my system of inductive logic a rule specifying the particular kind of properties for which inductive projection is permitted, the reply is: no additional rule is required; the definition of degree of confirmation is complete, and is sufficient to determine the kind of property for which projectability holds. (1947, p. 147)

But it remains to be established that the class of predicates that are "purely qualitative" in the intuitive sense (e.g., 'blue') is or ought to be identified with the class that underwrites the technical results on probabilistic projectability. Thus Goodman's original complaint stands, but a major proviso is needed. The "ingenious and valuable logico-mathematical apparatus" that Goodman thought to be so much idle machinery without a prior solution to the problem of projectability turns out to have a positive role in framing the problem.

As a postscript, I will mention one episode in a fascinating correspondence between Carnap and Hempel that took place during the period of 1946 to 1947, when Goodman introduced the grue problem.[13] At one point Hempel commented that Goodman's problem is closely related to the shopworn point that "any given finite evidence satisfies several incompatible general regularities, which lead to different predictions for future cases."[14] I, of course, agree with Hempel's remark, since it identifies Goodman's problem as an aspect of the general problem of underdetermination. But I want to note that if 'satisfies' is taken to mean confirms, then Hempel's remark is a strange one for him to have made, since one of the adequacy conditions he imposes on an account of qualitative confirmation is the consistency condition, requiring that the set of hypotheses confirmed by any self-consistent evidence must be mutually consistent (see chapter 3). If one is thinking of hypotheses that are confined to observational generalizations that can be stated in first-order predicate logic without function symbols (as Hempel was when writing "Studies in the Logic of Confirmation," 1945), then a consistent E will not "satisfy" incompatible hypotheses unless Goodmanized predicates are involved (a possibility Hempel did not envision when writing "Studies"). Indeed, we can use this fact to define the notion of a Goodmanized predicate, or rather, we can say of two predicates that each is Goodmanized relative to the other just in case the semantic rules of the language allow that there are individuals that simultaneously satisfy both predicates, whereas for other individuals to simultaneously satisfy both implies a contradiction. (Which member of the pair is "really" Goodmanized is, as Goodman has shown, a delicate matter to decide.) I think that the fact that Goodmanized predicates have to be employed in the limited setting presupposed by Hempel confirmation theory to produce examples of evidence that satisfies incompatible hypotheses is in part responsible both for the misimpression that Goodman's problem involves a trick and for the further misimpression that the problem is to be solved by a clever logical maneuver. When we move to a richer language, both misimpressions immediately disappear, for when hypotheses formulated in a theoretical vocabulary are allowed, it is easy to produce examples where the same evidence satisfies any number of incompatible hypotheses that do not involve any gerrymandered predicates. What is then called for is not a clever logical maneuver but a judgment as to which of the conforming hypotheses is better supported by the evidence. This is a problem for which Bayesianism seems tailor-made. In later chapters we will see how well the clothes fit.

Finally, I cannot resist speculating about how the later Carnap should have responded to Goodman's *Fact, Fiction, and Forecast*.[15] The Carnap I have in mind was tending in his later years more and more toward Bayesian personalism. This Carnap should have said to Goodman, "You applaud Hume's descriptive approach to the problem of induction, but you fault him for the inaccuracy of his description. I fault your approach in *Fact, Fiction, and Forecast* on the same grounds. You set for yourself the task of describing rules of projectability with which all (rational?) inductive agents operate. There are no rules of the kind you seek. Different agents project different predicates to different degrees, and the only constraints on the projections are those provided by Bayesian personalism." In chapter 6, I will examine the implications of such a position for the objectivity of scientific inference.

8 Novelty of Prediction and Severity of Test

In this chapter and the preceding one I have emphasized the merits of the Bayesian approach to confirmation. In the following chapters I will focus on what I take to be the most serious weaknesses of Bayesian methodology. In between these two extremes lies a murky area where it is hard to discern whether Bayesianism earns merits or demerits.

A prime instance of the murk stems from the widely held notion that novelty of predictions is an important factor in assessing the evidential support of theories. Many commentators have focused on what may be dubbed *temporal novelty*, and various remarks in the literature indicate that the following is taken to be a methodological truism about temporal novelty:

N1 Suppose that $T \models E$. If E was already known to be true prior to the articulation of T, then E does not confirm T.

Or more cautiously:

N2 Suppose that $T \models E_1$ and $T \models E_2$. If E_1 was known to be true prior to the articulation of T whereas E_2 became known only afterward, then E_2 confirms T more than does E_1.

Both (N1) and (N2) are subject to solid counterexamples drawn from the history of science, as will be discussed in detail in chapter 5. (For instance, take T to be Einstein's general theory of relativity [GTR], E_1 to be the data about Mercury's perihelion, and E_2 to be the data about the bending of star light during a solar eclipse. E_1 but not E_2 was known to be true prior to the articulation of GTR. But almost without exception, physicists say that E_1 affords GTR better confirmation than does E_2.) The fact that "old evidence" can give better confirmational value than "new evidence" poses a major problem for Bayesianism. In this section I will have to ignore as best I can this thorny problem.

John Worrall (1985, 1989) has championed the importance of novel predictions, but for him it is *use novelty* as opposed to temporal novelty that matters. In broad terms, his claim is the following:

N3 Suppose that $T_1 \models E$ and $T_2 \models E$. If E was used in constructing T_1 but not in constructing T_2, then T_2 receives more support from E than does T_1.

The Bayesian evaluation of (N3) begins with the observation that since both T_1 and T_2 are assumed to entail E, it follows that $\Pr(T_{1,2}/E) - \Pr(T_{1,2}) = \Pr(T_{1,2})[(1/\Pr(E)) - 1]$. So if 'support' in (N3) is interpreted as incremental confirmation (and if the problem of old evidence is ignored), (N3) is valid by Bayesian lights just in case $\Pr(T_2) > \Pr(T_1)$ whenever E was used in constructing T_1 but not T_2.

The notion that E was used in constructing T_1 is vague and ambiguous. (For example, did Einstein use the data concering Mercury's perihelion in constructing GTR? As will be seen in chapter 5, the answer is affirmative in the sense that on the way to the final version of his theory he rejected alternative theories because they failed to yield the correct perihelion prediction. On the other hand, he didn't explicitly use the perihelion data in constructing his field equations, and he had to do an elaborate calculation to convince himself that these equations did predict a shift of 43 seconds of arc per century.) However, there is a clear-cut class of cases, namely, those where T_1 comes from a more general theory T_1' by fitting T_1' to E, e.g., T_1' contains a free parameter whose value is fixed or at least circumscribed by E. A concrete example is supplied by the Brans and Dicke theory of gravitation. Until the value of the parameter ω is fixed, the theory yields no definite prediction about the three classical tests of GTR: the advance of the perihelion of Mercury, the bending of light, and the red shift. But the results of these three experiments place constraints on ω. Since these results were not used in constructing GTR—at least not in the parameter-fixing sense, since GTR contains no free parameters—it follows from (N3) that GTR should be better supported by the classical tests than the constrained Brans and Dicke theory. However, my informal survey has found that research workers in relativistic gravitational theory do not show much enthusiasm for this conclusion. On the Bayesian analysis, this is not surprising, since it not at all clear that GTR deserves a higher prior than the constrained Brans and Dicke theory.

What is Worrall's rationale for (N3)? He writes,

As for the question of *why* it should matter, once a theory has been produced, *how* it was produced, my answer in outline is this. Whether or not it can be given some further rationale, we *do* seem to regard the striking empirical success for a theory as telling us something about the theory's overall—what? Truth? Verisimilitude? Probable truth? General empirical adequacy? Closeness to a natural classification? Take your pick. The reasoning appears to be that it is unlikely that the theory would have got this phenomenon precisely right just "by chance," without, that is, the theory's somehow "reflecting the blueprint" of the Universe. The choice between the "chance" explanation and the "reflecting the blueprint" explanation of the theory's success is, however, exhaustive only if a third possibility has been ruled out—namely that the theory was engineered or "cooked up" to entail the phenomenon in question. In this latter case, the "success" of the theory clearly tells us nothing about the theory's likely fit with Nature, but only about its adaptability *and* the ingenuity of its proponents. (1989, p. 155)

The Bayesian need not be unsympathetic to these sentiments, for insofar as Worrall's rationale is a good one, it can be given a Bayesian reading, since information about the genesis of a theory is clearly relevant to the assessment of its prior probability. The rationale doesn't appear to be a good one in the clearest class of cases where E is used in the construction of T, the parameter-fixing cases. There may be other cases where the theory is cooked up to yield E in some sufficiently unappetizing way so that, as Worrall says, the success of the theory is more a reflection of the ingenuity of its proponents than of the likely fit of the theory with nature. But this is precisely what the Bayesian would construe as a plausibility argument that nudges lower the assignment of the prior probability to the theory.

Two rejoinders to the Bayesian assay of (N3) are worth mentioning. The first is that since the assignment of priors is a tricky and subjective business, the Bayesian reading of (N3) gives it a much less firm status than it was intended to have. On behalf of the Bayesians I reply that this is as it should be, since the implication of novelty for confirmation is a matter of considerable disagreement among both scientists and philosophers of science. (But, of course, the problem of priors cannot be swept under the rug. It will receive my full attention in chapter 6.) The second rejoinder is that the Bayesian interpretation of Worrall's rationale relies on a judgment of the efficacy of a plausibility argument, and on pain of circularity, the judgment cannot be analyzed in Bayesian terms. This objection was already considered in section 9 of chapter 2.

Presumably, the proponent of use novelty should also subscribe to the following:

N4   Suppose that $T \models E_1$ and $T \models E_2$. If E_1 but not E_2 is used in the construction of T, then T receives more support from E_2 than from E_1.

On the Bayesian analysis, (N4) is valid just in case $\Pr(E_2) < \Pr(E_1)$ whenever E_1 but not E_2 is used in the construction of T. Since this condition is patently absurd—why should the prior likelihood of the evidence depend upon whether it was used in constructing T?—the Bayesians and the proponents of (N4) are at loggerheads. I leave it to the reader to judge who the winner is by considering historical examples.

Deborah Mayo (1991) has argued that the concern with novelty is misplaced. What is really at issue, she maintains, is the severity of a test of a hypothesis, and the resort to novelty is but a ham-handed way of trying to guarantee severity. While I tend to agree with her diagnosis, I am concerned that her own criterion of severity is too demanding. According to her account, passing a test with outcome E counts as support for hypothesis H only if the test is *severe* in that the probability is high that such a passing result would not have occurred were E false. I take this subjunctive clause to imply that $Pr(E/\neg H \& K)$ is low (where K is the background knowledge). If this consequence obtained, it would be a beautiful result, since if H and $\neg H$ have comparable priors, it follows that outcome E gives H a high posterior probability (see chapter 7). The difficulty is that when high-level theoretical hypotheses are at issue, we are rarely in a position to justify a judgment to the effect that $Pr(E/\neg H \& K) \ll .5$. If we take H to be Einstein's general theory of relativity and E to be the outcome of the eclipse test, then in 1918 and 1919 physicists were in no position to be confident that the vast and then unexplored space of possible gravitational theories denoted by $\neg GTR$ does not contain alternatives to GTR that yield the same prediction for the bending of light as GTR. Today we know that $\neg GTR$ contains an endless string of such theories. The difficulties this makes for a Bayesian account of the confirmation of theories like GTR will be discussed in chapter 7.

9   Conclusion

Perhaps full-time Bayesians will not be satisfied with my efforts at defending their doctrine. But I trust that I have managed to reveal one of the undeniably impressive properties of Bayesianism: the more it is attacked, the stronger it gets, and the more interesting the objection, the more interesting the doctrine becomes. This feature, together with the positive successes of Bayesianism and the failures of alternative views, certainly justify giving Bayesianism pride of place among approaches to confirmation and scientific inference.

5 The Problem of Old Evidence

1 Old Evidence as a Challenge to Bayesian Confirmation Theory

One of the great virtues of Bayesian confirmation theory is its ability to pinpoint and explain the strengths and weaknesses of rival accounts, or so it was claimed in chapters 3 and 4. Recall, in particular, that I claimed that Bayesianism explains why the HD method of confirmation works as well as it does. To review, suppose that the confirmatory power of E for T is measured by $C(T, E) \equiv \Pr(T/E) - \Pr(T)$.[1] Suppose further that (1) $T \models E$ (the basic HD condition), (2) $1 > \Pr(T) > 0$, and (3) $1 > \Pr(E) > 0$. Then by Bayes's theorem, $C(T, E) > 0$, in consonance with the HD method. Furthermore,

$$C(T \ \& \ X, E)/C(T, E) = \Pr(T \ \& \ X)/\Pr(T) = \Pr(X/T),$$

which shows how tacking an irrelevant conjunct X onto T serves to reduce confirmatory power.

This display of virtue also serves to reveal an Achilles' heel of Bayesianism, or so Glymour (1980) has argued. To see the difficulty in concrete terms, take the time to be November 1915, when Einstein formulated the final version of his general theory of relativity (GTR) and when he first showed that this theory explains the heretofore anomalous advance of the perihelion of Mercury.[2] The nature of Mercury's perihelion had been the subject of intensive study by Le Verrier, Newcomb, and other astronomers, and so the relevant facts E were old evidence. In Bayesian terms, this means that for any agent who was conversant with the field and who operated according to the model of learning from experience by strict conditionalization (see chapter 2), $\Pr_{1915}(E) = 1$. Thus condition (3) above fails, with the results that $\Pr_{1915}(T/E) = \Pr_{1915}(T)$ and $C(T, E) = 0$ for any T and thus for GTR in particular, which seems to run counter to the generally accepted conclusion that the facts E did in November 1915 (and still do) provide strong confirmation of GTR. Indeed, in an exhaustive survey of the literature Brush (1989) found that with very few exceptions physicists have held that the perihelion phenomenon gives better confirmational value than either of the other two classical tests—the bending of light and the red shift—despite the fact that the former was old news while the latter two represented novel predictions.

Replacing the incremental with the absolute criterion of confirmation, according to which E confirms T at t just in case $\Pr_t(T/E) > r$ for some fixed $r > 0$, allows old evidence to have confirmational value. But the value of the absolute degree of confirmation may not capture the strength of the evidence, since if E' and E'' are both old at t, then $\Pr_t(T/E') = \Pr_t(T/E'') = \Pr_t(T)$.[3]

In section 2, I will examine some preliminary attempts to solve or dissolve the problem of old evidence. Although unconvincing, these attempts nevertheless serve the useful functions of distinguishing different versions of the problem and of pinpointing the version most worthy of attention. Sections 3 to 5 discuss an approach to this worthy version due to Garber (1983), Jeffrey (1983a), and Niiniluoto (1983). Section 6 takes up the neglected flip side of the old evidence problem: the problem of new theories. Section 7 summarizes some pessimistic conclusions for the prospects of Bayesian confirmation theory in the light of the old-evidence problem.

2 Preliminary Attempts to Solve or Dissolve the Old-Evidence Problem

Resorting to an objectivist interpretation of probability would help if it led to $\Pr(E) < 1$ for old evidence E. But insofar as I have a grasp of objective probability interpreted either as propensity or as relative frequency, it would seem that the objective probability is 1 (or else 0) for an anomalous perihelion advance of 43 seconds of arc per century, at least on the assumption that a deterministic mechanism is operating. This difficulty need not arise if objective probability means not propensity or frequency but the uniquely determined rational degree of belief. Thus Rosenkrantz (1983) recommends that we compute $\Pr(E)$ relative to a "considered partition" $H_1, H_2, \ldots, H_n$: $\Pr(E) = \sum_{i=1}^{n} \Pr(E/H_i) \times \Pr(H_i)$. He claims that unless E is a necessary truth, this sum will be less than 1 and will remain less than 1, since the likelihoods $\Pr(E/H_i)$ are "timeless relations." I have two difficulties with this tack. First, I am unpersuaded by attempts to objectify the assignments of prior probabilities.[4] Second, if $\Pr(\cdot)$ is interpreted as degree of belief, rational or otherwise, then it must be time-indexed, and $\Pr_{1915}(E)$ would seem to be 1.

Another way to try to resolve the problem of old evidence would be to insist upon using a conditional probability function $\Pr(\cdot/\cdot)$ that is defined even when the conditioning propositions have 0 unconditional probability (see appendix 1 to chapter 2). This allows one to adopt as a measure of evidential support $\hat{C}(T, E) \equiv \Pr(T/E) - \Pr(T/\neg E)$. When $0 > \Pr(T) > 1$,

$C(T, E) = (1 - \Pr(T))\hat{C}(T, E)$, so the new measure agrees qualitatively with the old as to positive and negative relevance. But the seeming advantage of the new measure is that in the case of old evidence ($\Pr(E) = 1$), when $C(T, E) = 0$, $\hat{C}(T, E)$ can be positive. Unfortunately, $\hat{C}(T, E)$ is not a suitable measure of confirmatory power. In the HD case ($T \models E$) with old evidence ($\Pr(E) = 1$), $\hat{C}(T, E) = \Pr(T)$, which means that all such evidence is counted as having the same confirmatory power.[5]

The original problem of old evidence would vanish for Bayesian personalists for whom $\Pr(E) \neq 1$, with Pr interpreted as personal degree of belief.[6] There are both historical and philosophical reasons for such a stance. In my running example, the literature of the period contained everything from 41″ to 45″ of arc per century as the value of the anomalous advance of Mercury's perihelion, and even the weaker proposition that the true value lies somewhere in this range was challenged by some astronomers and physicists.[7] Of course, if we push this line to its logical conclusion, we will eventually reach the position that no "thing language" proposition of the sort useful in confirming scientific theories is ever learned for certain, and the strict conditionalization model will collapse. Bayesians are hardly at a loss here, since Jeffrey (1983b) has proposed a replacement for strict conditionalization that allows for uncertain learning (see chapter 2).

However, denying that $\Pr(E) = 1$ only serves to trade one version of the old-evidence problem for another. Perhaps it was not certain in November 1915 that the true value of the anomalous advance was roughly 43″ of arc per century, but most members of the scientific community were pretty darn sure, e.g., $\Pr(E) = .999$. Assuming that Einstein's theory does entail E, we find that the confirmatory power $C(T, E)$ of E is $\Pr(T) \times .001/.999$, which is less than .001002. This is counterintuitive, since, to repeat, we want to say that the perihelion phenomenon did (and does) lend strong support to Einstein's theory.[8]

In what follows, then, I will work within the Bayesian personalist framework, using strict conditionalization to model learning from experience, and I will use $C(T, E)$ as the measure of confirmatory power. I will first consider the response that a proper use of this apparatus shows the old-evidence problem to be a pseudoproblem for logically omniscient Bayesian agents. Here logical omniscience involves two elements. The first (LO1) embodies the assumption that all the logical truths of the language L on which Pr is defined are transparent to the Bayesian agent. This assumption is codified in the basic axiom that if $\models X$ in L, then $\Pr(X) = 1$. Thus failure to accord maximal probability to logical truths of L leads to Dutch-book situations. The second element (LO2) involves the assumption that the agent is aware of every theory that belongs to the space of possibilities. In effect, when making the starting probability assignments $\Pr_{t_0}$, the agent formulates and considers every theory that can be stated in L. Now take a piece of empirical evidence E about which the agent was not certain *ab initio*, i.e., $1 > \Pr_{t_0}(E) > 0$, and suppose that the agent learns E between t_n and t_{n+1}. Then if $\Pr_{t_n}(T/E) > \Pr_{t_n}(T)$, a confirmational event takes place.[9] That event takes place only once, since on the strict conditionalization model, for any $m > n + 1$, $\Pr_{t_m}(E) = 1$. But once is enough, for at t_m we can still say that E is good evidence for T, since the history of the present probability function $\Pr_{t_m}$ contains the relevant sort of confirmational event.

This solution does not apply to real-world Bayesian agents who violate (LO2). In my running example, this includes the entire physics community in 1915, since Einstein's general theory was not formulated until the end of November of that year. Of course, if we could succeed in showing in Bayesian terms how GTR was confirmed for real-life scientists in 1915, then we could use the above strategy to cover post-1915 times.

As an aside it may be helpful here to refer to Eells's (1985) revealing classification of the problems of old evidence:

I. The problem of *old new evidence*: T was formulated before the discovery of E, but it is now later, and $\Pr(E) = 1$. So $\Pr(T/E) = \Pr(T)$.

II. The problem of *old evidence*: E was known before the formulation of T.
   A. The problem of *old old evidence*: It is now some time subsequent to the formulation of T.
      1. T was originally designed to explain E.
      2. T was not originally designed to explain E.
   B. The problem of *new old evidence*: It is now the time (or barely after the time) of the formulation of T.
      1. T was originally designed to explain E.
      2. T was not originally designed to explain E.

Taking into account confirmational histories and confirmational events seems to solve (I), and given a solution to (II.B), it also serves to solve (II.A). The remaining problem is that of new old evidence. The only places I differ with Eells are over cases (II.A.1) and (II.B.1), where Eells assumes that E cannot confirm T (see the discussion below in sections 5 and 6).

To return to the main discussion, Garber (1983), Jeffrey (1983a), and Niiniluoto (1983) have reacted to this version of the old-evidence problem by proposing to drop (LO1) as well as (LO2). Dropping (LO1) allows Bayesian agents to do logical and mathematical learning, and such learning, so they claim, can serve to boost the probability of the theory, as required by the incremental analysis of confirmation. What Einstein learned in November of 1915, so the story goes, was that his general theory entailed the heretofore anomalous perihelion advance, and conditionalization on that new knowledge was the relevant confirmational event.[10] I will examine this line of attack in detail in the following sections, but before doing so, I will comment briefly on another tack.

The alternative to Garber, Jeffrey, and Niiniluoto (GJN) is to demonstrate incremental confirmation for counterfactual degrees of belief, using degrees of belief the agent would have had if he hadn't known E prior to the formulation of T.[11] But as Chihara (1987), Eells (1985), Glymour (1980), van Fraassen (1988) and other commentators have objected, it is not evident that the relevant counterfactual degree of belief will be determinate or even that it will exist. In my historical example it is relevant that in 1907 Einstein wrote, "I am busy on a relativistic theory of the gravitational law with which I hope to account for the still unexplained secular change of the perihelion motion of Mercury. So far I have not managed to succeed" (Seelig 1956, p. 76). Thus it is not beyond the pale of plausibility that if Einstein hadn't known about the perihelion phenomenon, he wouldn't have formulated GTR. And if someone else had formulated the theory, Einstein might not have taken it seriously enough to assign it a nonzero prior, or he might not have understood it well enough to assign it any degree of belief at all. I will return to counterfactual degrees of belief in section 7.

3 Garber's Approach

To illustrate how logical omniscience (LO1) can be abandoned so as to make way for a more realistic Bayesianism, Garber (1983) begins with a language L in which distinct atomic sentences are treated as logically independent and in which the nonatomic sentences are all truth-functional compounds of the atomic ones. He then moves to a richer language L^* that contains the sentences of L and also new atomic sentences of the form $X \vdash Y$, where X and Y are sentences of L.[12] The symbol '$\vdash$' is a primitive connective of $L*$, but the aim is to interpret it as logicomathematical implication in whatever system of logic and mathematics is needed for the branch of science in question. Toward this end, Garber requires that under the Pr function, '$\vdash$' behaves as if it obeys modus ponens:

$$Pr((X \vdash Y) \& X) = Pr((X \vdash Y) \& X \& Y) \qquad (G)$$

Garber then shows that learning that $T \vdash E$ can serve to confirm T. More specifically, he shows that there is a probability Pr defined on the sentences of $L*$ and satisfying (G) such that $0 < Pr(A \vdash B) < 1$ whenever A and $\neg B$ are not both tautologies of L. It may therefore be that $Pr(A/A \vdash B) > Pr(A)$. We may wish to add the further constraint that $Pr(A \vdash B) = 1$ whenever $A \to B$ is a tautology of L. But since Einstein's GTR does not truth-functionally entail the perihelion advance evidence E, it is consistent with this constraint to set $Pr(GTR \vdash E) < 1$ and thus to have $Pr(GTR/GTR \vdash E) > Pr(GTR)$.

Three criticisms have been brought against this approach. The first is that Garber has only shown that a solution to the problem of old evidence is possible within the framework of the Bayesian strict conditionalization model and not that a solution of this form actually applies to the historical case at issue. To complete the solution for the case of the perihelion of Mercury, it would have to be demonstrated that there is a plausible set of constraints that Einstein's degrees of belief did or should have satisfied and that guarantee that his learning that GTR $\vdash$ E served to boost his degree of belief in GTR. This hiatus was addressed by Jeffrey (1983a), whose attempt to fill the gap will be examined in detail in the next section.

A second criticism of Garber's approach derives from the observation that his approach requires logical omniscience (LO1) with respect to the truth-functional logic of $L*$ but not with respect to predicate logic, arithmetic, or calculus. But demanding knowledge of very complicated truth functional implications can be even more unrealistic than demanding knowledge of simple truths of arithmetic or calculus. This leads Eells (1985) to propose that a truly realistic version of Bayesianism should set the standard of what logicomathematical truths the Bayesian agent is supposed to know in terms of complexity. I agree with Eells, but for present purposes it suffices to stick to Garber's preliminary version of (not thoroughly) humanized Bayesianism.

A third, and I believe unfair, category of criticism has been leveled in both the published literature and in informal discussions. To put it at its unfairest, the charge would go thus. To apply Garber's formalism to the problem of old evidence assumes that '$\vdash$' has been identified as the appropriate form of logicomathematical implication. But constraint (G) does not suffice for this identification, since other relations also satisfy (G). And if conditions sufficient to pin down the intended interpretation of '$\vdash$' are added to (G), there is no guarantee that Garber's demonstration of how learning $T \vdash E$ can serve to boost the probability of T will remain valid.

The response I propose on Garber's behalf is that (G) is not supposed to fix the interpretation of '$\vdash$'. The interpretation is fixed extrasystematically, e.g., by the intention of the agent to use '$\vdash$' to mean implication in some logicomathematical system. If it is then asked why (G) was imposed in the first place, two reasons can be given. It is important to demonstrate that Pr assignments can be made to reflect various constraints that '$\vdash$' ought to satisfy if it is interpreted extrasystematically as logicomathematical implication. And also, (G) plays an important role in proving that in actual historical cases $\Pr(T/T \vdash E) > \Pr(T)$, as will be seen below in section 4.

Van Fraassen (1988) is unconvinced that Pr assignments that resolve the old evidence problem can be made to conform to constraints appropriate to '$\vdash$' taken as a form of logicomathematical implication. In particular, he proposes that the form of a plausible constraint for conditional proof is the following:

If $\Pr'(X \,\&\, Y) = \Pr'(X)$ for all $\Pr'$ in $\mathscr{C}(\Pr)$, then $\Pr(X \vdash Y) = 1$.   (vF)

Here $\mathscr{C}(\Pr)$ is the class of alternatives to Pr that allow for the generalizations involved in conditional proof. Van Fraassen then poses the following dilemma for Garber. Suppose first that $\mathscr{C}(\Pr)$ consists of all probability functions that can be generated from Pr by strict conditionalization or by Jeffrey conditionalization or, more generally, by any shift in degrees of belief that does not change zero probabilities into nonzero probabilities. (In the jargon of probability, $\Pr'$ must be absolutely continuous with respect to Pr.) Then it follows from (vF) that

$$\Pr(X \vdash Y) = \Pr((X \vdash Y) \,\&\, (X \to Y))$$

and that

$$\Pr(X \to Y) = \Pr((X \to Y) \,\&\, (X \vdash Y)).$$

So $X \vdash Y$ is probabilistically indistinguishable from material implication. Thus if Y is old news, then so is $X \vdash Y$. On the other hand, to take $\mathscr{C}(\Pr)$

to include probability functions not absolutely continuous with respect to Pr is to consider priors we might have had if we didn't have the old evidence. But that is to enter the mire of counterfactual beliefs that I decided above in section 2 must be avoided.

On behalf of Garber, I propose to escape the dilemma by rejecting (vF) as a suitable means of reflecting conditional proof. Indeed, one can hold that no condition like (vF) is needed to reflect the successful application of a proof strategy, whether conditional proof or otherwise. Rather, the success is expressed in the Bayesian learning model. Thus, suppose that at time t_n the Bayesian agent shows that X implies Y by means of a conditional proof strategy, by a *reductio* strategy, by deriving Y from X and the accepted axioms by means of the accepted inference rules, or by whatever proof strategy is allowed in the relevant logic. Then the agent has learned that $X \vdash Y$, and so on the strict conditionalization model, $\mathrm{Pr}_{t_{n+1}}(X \vdash Y) = 1$.

My own concern about Garber's system lies not so much with qualms about lurking inconsistencies as with doubts about its relevance to the old-evidence problem. The axiom of probability requiring that $\mathrm{Pr}(X) = 1$ if $\models X$ is not contradicted in Garber's system by a value for $\mathrm{Pr}(\mathrm{GTR} \vdash E)$ lying strictly between 0 and 1, since in this system neither $\models (\mathrm{GTR} \vdash E)$ nor $\models \neg (\mathrm{GTR} \vdash E)$. For in Garber's system a possible world is given by an assignment of truth values to the atomic sentences; thus $\mathrm{GTR} \vdash E$ is true in some of the possible worlds and false in others. But then it is hard to see how the formal result that $\mathrm{Pr}(\mathrm{GTR}/\mathrm{GTR} \vdash E) > \mathrm{Pr}(\mathrm{GTR})$ bears on the idea that my learning that the theory entails E (in predicate logic or second-order logic or whatever) boosts my degree of belief in GTR, for the formal result holds only for a semantics that masks what $\mathrm{GTR} \vdash E$ is supposed to mean. In what follows, I waive this qualm because I think that the entire approach is beset by more fundamental difficulties.

4   Jeffrey's Demonstration

Let us now consider Jeffrey's attempt to show how learning that $T \vdash E$ can serve to boost the probability of T. Suppose the following:

$$\mathrm{Pr}(E) = 1 \tag{J1.a}$$

$$1 > \mathrm{Pr}(T) > 0 \tag{J1.b}$$

$$1 > \Pr(T \vdash E) > 0, \qquad 1 > \Pr(T \vdash \neg E) > 0 \tag{J2.a}$$

$$\Pr((T \vdash E) \,\&\, (T \vdash \neg E)) = 0 \tag{J2.b}$$

$$\Pr(T/(T \vdash E) \vee (T \vdash \neg E)) \geqslant \Pr(T) \tag{J3}$$

$$\Pr(T \,\&\, (T \vdash \neg E)) = \Pr(T \,\&\, (T \vdash \neg E) \,\&\, \neg E) \tag{J4}$$

Then $\Pr(T/T \vdash E) > \Pr(T)$.

Proof

$\Pr(T/(T \vdash E) \vee (T \vdash \neg E))$

$$= \frac{\Pr(T \,\&\, (T \vdash E)) + \Pr(T \,\&\, (T \vdash \neg E)) - \Pr(T \,\&\, (T \vdash E) \,\&\, (T \vdash \neg E))}{\Pr(T \vdash E) + \Pr(T \vdash \neg E) - \Pr((T \vdash E) \,\&\, (T \vdash \neg E))}$$

$$= \frac{\Pr(T/T \vdash E)}{1 + [\Pr(T \vdash \neg E)/\Pr(T \vdash E)]}.$$

The first equality follows by the definition of conditional probability and the standard axioms of probability. The second equality follows from the first, since by (J1.a) and (J.4), the second and third terms in the numerator on the right-hand side are 0, and by (J2.b), the third term in the denominator is 0. Then by (J2.a), the right hand side of the second equality is less than $\Pr(T/T \vdash E)$. But by (J3), the right-hand side of the second equality is greater than or equal to $\Pr(T)$, which gives the desired result.

Conditions (J1) and (J2.a) follow from the meaning of the problem of old evidence. Condition (J4) is just an application of Garber's (G). The crucial condition is (J3), which says in application to my running example that in November of 1915 Einstein's degree of belief in GTR before learning that it entailed the missing 43″ was less than or equal to his conditional degree of belief in the theory, given that the theory implies a definite result about the perihelion advance. Eells (1985) demurs that (J3) is suspect. For example, the above demonstration shows that in the presence of the other conditions (J3) leads to the result that if $\Pr(T \vdash E) = \Pr(T \vdash \neg E)$, then $\Pr(T/T \vdash E) \geqslant 2\Pr(T)$. This means that the prior probability of T cannot be greater than .5. And as $\Pr(T)$ approaches .5, $\Pr(T/T \vdash E)$ approaches 1, a wholly implausible result in the actual historical case at issue. I would add that (J2.b) is also suspect, for if we are supposed to be imagining humanized, nonlogically omniscient agents, it is unreasonable for them to be certain that a new and complicated theory is internally consistent.

Such difficulties do not lead Eells to reject the GJN approach. His stance is that the Bayesian can explicate "$T \vdash E$ confirms T" as $\Pr(T/T \vdash E) > \Pr(T)$ "without expecting there to be any single formal kind of justification … for exactly the cases in which $T \vdash E$ *should* be taken as confirming T" (1985, p. 299). While it is a fair comment that a single formal justification cannot be expected to cover all the cases, the GJN approach loses its interest if it cannot be shown that increases in probability will take place in an interesting range of cases. (Suppose that in the original setting, where the problem of old evidence is neglected, all the Bayesians could say is that $\Pr(T/E) > \Pr(T)$ happens when it happens. Then I would suggest that the number of adherents to Bayesian confirmation theory would dwindle. Fortunately, we can demonstrate that the inequality holds when various relations between T and E obtain and that these relations cover many of the cases where confirmation intuitively takes place.) It is to this question that I now turn.

As an alternative to Jeffrey's demonstration, consider the following:

$$\Pr(E) = 1 \tag{A1.a}$$

$$1 > \Pr(T) > 0 \tag{A1.b}$$

$$1 > \Pr(T \vdash E) > 0 \tag{A2}$$

$$\Pr((T \vdash E) \vee (T \vdash \neg E)) = 1 \tag{A3}$$

$$\Pr(T \,\&\, (T \vdash \neg E)) = \Pr(T \,\&\, (T \vdash \neg E) \,\&\, \neg E) \tag{A4}$$

Then $\Pr(T/T \vdash E) > \Pr(T)$.

Proof

$$\Pr(T) = \Pr(T \,\&\, [(T \vdash E) \vee (T \vdash \neg E)])$$

$$= \Pr(T \,\&\, (T \vdash E)) + \Pr(T \,\&\, (T \vdash \neg E))$$

$$\quad - \Pr(T \,\&\, (T \vdash E) \,\&\, (T \vdash \neg E))$$

$$= \Pr(T \,\&\, (T \vdash E)).$$

The first equality holds in virtue of (A3). The second follows from the first by the addition axiom. And the third follows in virtue of (A1.a) and (A4). Thus

$$\Pr(T/T \vdash E) - \Pr(T) = \frac{\Pr(T)}{\Pr(T \vdash E)}[1 - \Pr(T \vdash E)],$$

which, by (A1.b) and (A2), is greater than 0.

Since (A3) is stronger than Jeffrey's (J3), it might seem that this second derivation cannot be an improvement on the first. Note, however, that the present derivation did not have to rely on the suspect assumption that

$$\Pr((T \vdash E) \ \& \ (T \vdash \neg E)) = 0.$$

Yet it would also seem that this approach is also subject to Eells's type of objection, since it follows that $\Pr(T/T \vdash E) = \Pr(T)/\Pr(T \vdash E)$, which means that if $\Pr(T) = \Pr(T \vdash E)$, then $\Pr(T/T \vdash E) = 1$. But the result that $\Pr(T \ \& \ (T \vdash E)) = \Pr(T)$ says that the agent is certain that if T is true then $T \vdash E$. Since $T \vdash E$ is (probabilistically) a consequence of T, $\Pr(T) \leqslant \Pr(T \vdash E)$, with strict inequality typically holding. Thus, the analogue of Eells's objection for the present derivation lacks bite.

Alas, the problem of old evidence in the perihelion case remains unresolved. The meaty condition (A3) says that upon writing down his theory, Einstein was certain that it implied a definite result about the advance of the perihelion of Mercury. But the historical evidence goes against this supposition. Indeed, Einstein's published paper on the perihelion anomaly contained an incomplete explanation, since, as he himself noted, he had no proof that the solution of the field equations he used to calculate the perihelion was the unique solution for the relevant set of boundary conditions.

Assume that Einstein's degree of belief in GTR, conditional on the theory's giving the correct prediction for the perihelion of Mercury, was greater than his degree of belief in the theory, conditional on the theory's giving no definite prediction about the perihelion. If we let $T \vdash N$ stand for $\neg(T \vdash E) \ \& \ \neg(T \vdash \neg E)$, the assumption amounts to replacing (A3) with

$$\Pr(T/T \vdash E) > \Pr(T/T \vdash N). \tag{A3$'$}$$

Then (A1), (A2), (A3$'$), and (A4) together imply that $\Pr(T/T \vdash E) > \Pr(T)$.

Proof

$$\Pr(T) = \Pr(T/T \vdash E) \times \Pr(T \vdash E) + \Pr(T/T \vdash \neg E) \times \Pr(T \vdash \neg E)$$

$$+ \Pr(T/T \vdash N) \times \Pr(T \vdash N).$$

By (A1.a) and (A4), the second term on the right-hand side is 0. And since $\Pr(T \vdash E) + \Pr(T \vdash N) \leqslant 1$, $\Pr(T \vdash E) < 1$ (by (A2)), and $\Pr(T/T \vdash E) > \Pr(T/T \vdash N)$ (by (A3′)), the equality cannot hold if $\Pr(T) \geqslant \Pr(T/T \vdash E)$.

While the doubt left by Eells's analysis may not have been completely resolved in favor of the GJN approach, I think enough has been said to make it plausible that in an interesting range of cases, learning $T \vdash E$ can serve to boost confidence in T.

5   The Inadequacy of the Garber, Jeffrey, and Niiniluoto Solution

For those Bayesians who have been persuaded by Garber, Jeffrey, and Niiniluoto of the need to humanize their doctrine, the way is now open to search through the Einstein archives for evidence that in November of 1915 Einstein's beliefs conformed to (J1) through (J4) or to one of the alternative schemes (A1) through (A4) or (A1), (A2), (A3′), (A4). Suppose that the findings are positive (alternatively, negative). Would the problem of old evidence with respect to GTR and the perihelion of Mercury have thereby been shown to have a positive (alternatively, negative) solution? Not at all. The original question was whether the *astronomical data E* confirmed GTR (for Einstein, if you like). Garber, Jeffrey, and Niiniluoto replace this question with the question of whether Einstein's learning that $T \vdash E$ raised his confidence in the theory. Not only are the two questions not semantically equivalent; they are not even extensionally equivalent. We can say without a shadow of a doubt that for Einstein E did confirm T. But we have to be prepared for the archival finding that the conditions needed to prove that $\Pr(T/T \vdash E) > \Pr(T)$ fail for him.

The point becomes clearer when we shift from Einstein to others. We now want to say that the perihelion phenomenon was and is good evidence for Einstein's theory. But along with most students of general relativity, the first thing we may have learned about the theory, even before hearing any details of the theory itself, was that it explains the perihelion advance. So there never was a time for us when $\Pr(T \vdash E) < 1$.

Moreover, even if the two questions

Does E confirm T for person P?

Does learning $T \vdash E$ increase P's degree of belief in T?

should stand or fall together, there is no guarantee that the strength of confirmation afforded by E is accurately measured by the boost given to degrees of belief by learning $T \vdash E$. This matter is connected to the issue of whether E can confirm a theory designed to explain E.[13] It will be helpful here to distinguish three senses in which person P might be said to have designed T so as to explain E.

1. When P created T, he was motivated by a desire to explain E.
2. Before settling on T, P examined and rejected alternative theories that failed to explain E.
3. In arriving at T, P went through an explicit chain of reasoning that started with E and worked back to T.

As we move from (1) to (3), it becomes less and less surprising to P that $T \vdash E$, and therefore P's learning that $T \vdash E$ gives a smaller and smaller boost to his degree of belief in T. We have already seen that Einstein satisfied (1). That he satisfied (2) is indicated by the fact that he wrote to Sommerfeld in November of 1915 that one of the reasons he abandoned a previous theory, constructed with the help of his friend Marcel Grossmann, was that it yielded an advance of only 18″ per century for Mercury's perihelion.[14] But this piece of personal history does not seem to have diminished the confirmational value of E, as opposed to $T \vdash E$, for Einstein. Nor would the discovery that Einstein also satisfied (3) show that E had no confirmational value for Einstein or his fellow scientists.[15]

Substituting a tractable problem for an intractable one is a time-honored tactic. The tactic is fruitful if the solution to the tractable problem illuminates the original problem. In this case, however, the solution given by GJN to the Bayesian learning problem for humanized agents fails to speak to the original problem. Further, the so-far intractable part of the problem of old evidence is just as much a problem of new theories as of old evidence. How probability is to be assigned to the newly minted theory is a question that must be answered before we can begin to worry about whether and how the probability of T is boosted by E, by $T \vdash E$, or by whatever. The problem of new theories will be touched upon in section 6 and discussed in more detail in chapter 8. But before I close this section, it will be helpful to sketch a non-Bayesian account of why the perihelion data does constitute good confirmation of GTR.

Such an account could appeal to at least four facts: (1) that GTR yields the exact value of the anomalous advance, (2) that it does so without the help of any adjustable parameters, (3) that the perihelion phenomenon provides a good bootstrap test of GTR, and (4) that dozens of attempts were made within both classical and special relativistic physics to resolve the anomaly, all of which failed.[16] By way of explaining (3), assume that the exterior field of the sun is stationary and spherically symmetric. Then the line element can be written as

$$ds^2 = [1 - (\alpha m/r) + (2\beta m^2/r^2) + \cdots]dt^2$$
$$ - [1 + (2\gamma m/r) + \cdots](dx^2 + dy^2 + dz^2),$$

where m is the mass of the sun, $r^2 = x^2 + y^2 + z^2$, and α, β, γ are undetermined parameters. Einstein's GTR requires that $\alpha = \beta = \gamma = 1$. The first-order red shift depends only on α, while the bending of light depends only on α and γ. By contrast, the advance of the perihelion of a planet depends upon all three parameters, and so the perihelion data helps to pin down a parameter left undetermined by the other two classical tests.

The point here is that without doing any Bayesian calculations and without solving the Bayesian problem of old evidence, we can recognize on independent grounds the confirmational virtues of the perihelion data. Of course, there is nothing to block Bayesians from taking into account the factors enumerated above. But how these factors can be made part of Bayesian calculations in the context of old evidence remains to be seen.

6   New Theories and Doubly Counting Evidence

Despite my rejection of the GJN substitution move, I agree with the main thrust of their humanized Bayesianism: namely, a realistic theory of confirmation must take into account nonempirical learning. But while Garber, Jeffrey, and Niiniluoto address the learning of logicomathematical facts, they, like most of Bayesian authors, are silent about the learning of new theories, despite the obvious importance of such learning for an understanding of scientific change.[17] Indeed, I would venture that the problem of new theories presents both a more interesting challenge and a more interesting opportunity for Bayesians than does the original problem of old evidence.

In Bayesian terms, the introduction of new theories causes a humanized Bayesian agent (who fails (LO2)) to shift from a probability function Pr, operative before the introduction, to a new function Pr', operative after the introduction. And typically, Pr' is not derived from Pr by any straightforward conditionalization process. How this transition is or ought to be managed is a matter that I will take up in chapter 8. For present purposes, the details of how Pr' is generated are irrelevant.

The problem of new theories presents the opportunity to further explore the slogan that a theory T is not confirmed by evidence E that T was designed to explain. Suppose that E, whether fresh or stale, leads to the proposal of a new theory T, and suppose that this new T is assigned a nonzero probability relative to the new Pr' function. Then since the Pr' assignments were made in light of E, it would seem to be double counting the evidence to take it to confirm T in the sense of raising the Pr' probability of T.[18] That is the kernel of truth in the slogan. (Of course, if the agent fails logical omniscience (LO1) his assignment Pr'(T) may not accurately reflect the evidential import of E, for he may fail to know that $T \vdash E$, and upon learning the implication, he may change his degree of belief in T à la Garber and Jeffrey. But this does not undercut the prohibition against doubly counting evidence.)

It is worth noting that, looked at from the perspective of new theories, the problem of old evidence is not a problem at all but merely an application of the methodological truism that evidence should not be doubly counted. But looked at from the ex post facto perspective, the problem of old evidence is a real problem, since we want to affirm that E does after all confirm or support T.

7   Conclusion: A Pessimistic Resolution of the Old-Evidence Problem

The recognition that the interesting residual problem of old evidence arises from the problem of new theories is important, for it automatically undercuts some of the proposed treatments of old evidence. Suppose that the problem of a new theory has been resolved in that in reaction to the introduction of T the Bayesian agent chooses, in some appropriate way, a new probability function Pr' such that Pr'$(T) > 0$. In this setting, we cannot follow Howson's (1984, 1985) prescription for resolving the old-evidence problem by computing the difference between what the agent's degree of belief in T would have been if his total knowledge at the time T was introduced had been $K - \{E\}$, and what his degree of belief in T would have been were he subsequently to come to learn E.[19] At least we cannot take this computation to be given by the comparison of $\mathrm{Pr}'(T/K)$ and $\mathrm{Pr}'(T/K - \{E\})$, since both of these probabilities are equal to $\mathrm{Pr}'(T)$, unless the very introduction of T caused the agent to become uncertain about what he previously regarded as certain and accordingly to assign $\mathrm{Pr}'(K) < 1$. Howson's prescription is relevant for Bayesian agents who are logically omniscient in sense (LO2) and who change their belief functions only by conditionalization. But it is not a prescription that will cure the tough version of the old-evidence problem for agents who fail (LO2) and who resort to non-Bayesian shifts in their belief functions when new theories are introduced.

There seem to me to be only two ways to deal with the residual problem of old evidence. The first is to ignore it in favor of some other problem. That, in effect, is the route pioneered in the GJN approach. A perhaps better motivation for going this route derives from the view that ultimately our goal in scientific enquiry is to choose among competing theories, and for that choice, what matters are the relative values of the probabilities of the theories conditional on the total evidence, old as well as new (see, however, chapter 7).

But if the problem is not to be ignored, the only way to come to grips with it is to go to the source of the problem: the failure of (LO2). There are in turn two strategies for coping with the failure of (LO2), both involving counterfactual degrees of belief based not just on counterfactual evidence sets but on counterfactual probability functions. One version, already mentioned in section 2 above, imagines that the agent is empirically deficient as well as logically deficient. It imagines that the agent didn't know E and asks what, in these circumstances, the agent's degree of belief in T would have been when T was introduced, and then it compares that number with what the agent's subsequent degree of belief in T would have been had he then learned E. The computation is thus done using not the probability function Pr' actually adopted by the agent upon the introduction of T but a hypothetical function. The other version imagines what the agent's degree of belief in T would have been *ab initio* if he were not logically deficient but were a superhuman calculator satisfying (LO1) and

(LO2), and then it compares this number with the degree of belief this supercalculator assigns after learning E. This calculation involves a hyper-hypothetical probability function. I have no doubt that counterfactual enthusiasts can concoct ways to get numbers out of one or both of these scenarios. But what has to be demonstrated before the problem of old evidence is solved is that the numbers match our firm and shared judgments of the confirmational values of various pieces of old evidence. It would be quite surprising if such a demonstration could be given, since the counterfactual probabilities and thus the counterfactual incremental boosts in confirmation will vary greatly from one person to another. Hope springs eternal, but even if the hope is realized the Bayesian account of confirmation retains a black eye for being forced to adopt such a complicated and dubious means of accommodating such a simple and common phenomenon in scientific inference.

6 The Rationality and Objectivity of Scientific Inference

1 Introduction

The rationality and objectivity of scientific inference is a topic that has been much on the minds of philosophers of science over the past few decades. But all too often these minds have been abuzz with incommensurability, relativism, duck-rabbit gestalt switches, the theory-ladenness of observation, and the like. I will have more to say about these matters in chapter 8, but for present purposes I will assume that we have all taken a magic pill that has cured this particular buzz. While this strategy will be unappealing to some readers, I think it serves to make the problem of rationality and objectivity more interesting, since, if I am right, the problem arises even if the "new fuzziness" (as Clark Glymour has dubbed it) is set aside.

For the purest of the Bayesian personalists, the constraints of rationality begin and end with the axioms of probability. Less extreme personalists want to impose the conditionalization model for learning from experience and perhaps some form of David Lewis's principal principle (chapter 2). If these procedural constraints are satisfied, will the resulting degrees of belief count as rational? And if not—if there is a mismatch between the rationality of the result and the alleged rationality of the procedure—can we properly say that the procedure is fully rational? To put the matter in concrete terms, if in the face of the currently available evidence you assign a high degree of belief to the propositions that Velikovsky's *Worlds in Collision* scenario is basically correct, that there are canals on Mars, that the earth is flat, etc., you will rightly be labeled as having an irrational belief system. And if you arrived at your present beliefs within the framework of Bayesian personalism, then the temptation is to say that at worst there is something rotten at the core of Bayesian personalism and at best there is an essential incompleteness in its account of procedural rationality.

Here the issue of rationality merges with that of objectivity. We use epithets like 'irrational' and 'crackpot' in the case I just described and in other cases where there is objectivity of opinion in the sense of a tight intersubjective agreement in the relevant scientific community and where the stigmatized person has opinions that differ radically from the consensus. Such objectivity exists in science not only concerning the roundness of the earth but for more theoretically interesting propositions as well, e.g., that matter has an atomic structure, that space and time are relativistic rather than absolute, etc. One's expectation, or hope if you will, is that the explanation of the intersubjective agreement on such matters is not merely historical or sociological but has a justificatory character—otherwise labeling as 'irrational' or a 'crackpot' those whose opinions depart radically from the consensus would be unfounded.

This sort of talk has traditionally been thought to presuppose a metaphysical thesis of objectivity to the effect that the object of scientific inquiry, "the world," exists independently of knowing subjects and that it makes sense to say of our opinions that either they match or fail to match this objective reality. In recent years the social constructivists have ridiculed this thesis as an outdated piece of ideology. I will not dignify constructivism with a response but will instead concentrate on what I take to be a more interesting but unmet challenge to the objectivity of scientific inference.

The popular image of science would have it that science provides us with a methodology for generating objective opinion: apply the scientific method faithfully and long enough and eventually it will produce certitudes that match reality, and before certainty is reached, the faithful use of the method by all the members of the scientific community guarantees objectivity qua intersubjective agreement on degrees of belief. These degrees of belief are rational because they are produced by an objective method of inquiry: it is value-free and presupposition-free, it is evidence-driven, and it sanctions no inference not strictly warranted by the evidence.[1] This straw man, or as I would prefer to say, this wish list, has been criticized on the grounds that nothing remotely like it can be instantiated because of incommensurability, the theory-ladenness of observation, and the like. Again I am postponing this challenge until chapter 8 to take up a more fundamental challenge: even if we leave aside incommensurability and its fellow travelers, there is still no obvious candidate for this objective method of inquiry. Certainly Bayesianism doesn't qualify, since it not only allows but requires presuppositions in the form of prior probabilities.

Failure to fulfill the pop-science specifications for an objective method of inquiry does not mean that Bayesianism cannot deliver on objectivity. We will see that the long-run use of the Bayesian method does produce certainties that "almost surely" match reality. At least this is so for observational hypotheses; theoretical hypotheses are another matter, one that requires careful discussion. Less satisfactory are attempts to ground objectivity as intersubjective agreement for opinions that fall short of certainty. Two of the most widely used attempts to provide the grounding within the

Bayesian framework are (1) constraining priors and (2) the washing out of priors. Both attempts, I will argue, have only limited success. Several other alternatives present themselves: (3) definitional solution, (4) socialism, (5) evolutionary solution, (6) modest but realistic solutions, (7) non-Bayesian solutions, and (8) retrenchment. Each of these alternatives will also be found wanting.

2 Constraining Priors

The first strategy for grounding objectivity as intersubjective agreement proceeds by supplementing the personalist form of Bayesianism by adding constraints on prior probabilities. The literature contains numerous and telling objections to various attempts to implement this strategy. I will not review this literature here but will simply indicate briefly why such attempts are unworkable and why they would not solve the problem of rationality and objectivity even if they were workable.

There are two reasons why such principles are unworkable. The first is that different applications of these principles are possible, and the different applications can yield conflicting results. This phenomenon was illustrated in chapter 1 for Bayes's particular application of the principle of insufficient reason. Nor is the problem of choosing the "correct" application much easier than the general problem of deciding what to believe. This is again illustrated by Bayes's case, since the application favored by Bayes leads to an inductivist probability with a high rate of learning from experience, whereas another application led to a noninductivist probability with no learning from experience. What holds for Bayes's rule for assigning priors holds quite generally: there are different ways of conceptualizing an inference problem, and the application of the rule to the different conceptualizations leads to different results. The problem of choosing among the different results is no less difficult than the original problem of assigning priors.

Even if there were no ambiguity in the conditions of application of these principles, there would remain the problem that the conditions are rarely satisfied in real-life cases. Recall that Bayes assumed a condition of complete ignorance regarding the unknown event. If we ignore the potential ambiguities in this notion, the point is that such a condition will be realized only in never-never-land Bayesianism, where an agent begins as a *tabula rasa*, chooses her priors, and forever after changes her probabilities only by conditionalization. A more realistic Bayesianism would recognize the local and episodic character of problem solving. In Bayesian terms, we use different probability functions for different problem-solving contexts, and within a context we may change probabilities not by conditionalization but by some more radical means.[2] Thus, far from being a *tabula rasa*, the typical scientist comes burdened with a wealth of information in trying to make what the Bayesian would describe as decisions about prior probabilities. E. T. Jaynes's modern version of the principle of indifference tries to take into account some of this information, since it enjoins us to maximize a quantity he calls "entropy" subject to known constraints that can be expressed in terms of moments of the probability distribution.[3] But only a small part of prior information can be expressed in these terms.

How, then, is prior information taken into account? What one finds running through the scientific literature are plausibility arguments. In Bayesian terms, these arguments are designed to persuade us to assign high priors to some alternatives and low priors to others.[4] Part of what it means to be an "expert" in a field is to possess the ability to recognize when such persuasions are good and when they are not. But it is highly doubtful that this ability can be codified in simple formal rules. And even if it could, why is or should only expert opinion be tolerated?

Suppose now for the sake of argument that there are workable rules for assigning priors. There are still two reasons why these rules will not suffice to explain objectivity. First, the explanation would have the sought-after justificatory character only if the rules in question were accepted as norms of rational behavior. But their normative status is highly controversial; indeed, these rules are either explicitly rejected or else ignored by a large segment of the Bayesian camp. Second, even if these rules were uniformly accepted, they would not be sufficient to explain objectivity unless they sufficed to fix the likelihoods $\Pr(E/H_i)$ needed to implement the right-hand side of Bayes's theorem in the form (2.2). But these rules are typically intended to fix the prior probabilities on a partition $\{H_i\}$ of hypotheses and are not intended to apply to partitions such as $\{E \mathbin{\&} H_i\}$, where E is a possible data report.[5] There are, of course, cases where the likelihoods do have an objective status. The HD case, where for each i either $H_i \mathbin{\&} K \models E$ or $H_i \mathbin{\&} K \models \neg E$, is one such. Another obtains when all the Bayesian agents agree on a statistical model for a chance experiment, E reports outcomes of the experiment, the H_i are alternative hypotheses about the objective-probability parameters of the chance setup, and Lewis's principal principle applies. But these cases hardly exhaust the domain that would have to be covered by an adequate theory of scientific inference. Consider, for example, as astronomers of the seventeenth century were forced to, what probability should be assigned to stellar parallax of various magnitudes on the assumptions of a Copernican cosmology and the then accepted background knowledge, which contained scanty and uncertain information about the distance between the earth and the fixed stars.[6]

3   The Washing Out of Priors: Some Bayesian Folklore

Many Bayesians analyze objectivity in terms of the washing out of priors. Thus Adrian F. M. Smith writes,

> I personally am only able to make sense of the concept [scientific objectivity] in the context of a Bayesian philosophy that predisposes one to seek to report, openly and accessibly, a rich range of possible belief mappings induced by a given data set, the range being chosen to reflect and potentially to challenge the initial perceptions of a broad class of interested parties. If a fairly sharp consensus of views emerges from a rather wide spread of initial opinions, then, and only then, might it be meaningful to refer to "objectivity." (1986, p. 10)

The implication of Smith's suggestion is that even if there were workable principles for constraining priors, it would be a mistake to impose them. It is a fact of life that scientists start with different opinions. To try to quash this fact is to miss the essence of scientific objectivity: the emergence of an evidence-driven consensus from widely differing initial opinions.

It is part of the Bayesian folklore that the emergence of such a consensus is routine. Differences in prior probabilities do not matter much, at least not in the long run; for (the story goes) as more and more evidence accumulates, these differences wash out in the sense that the posterior probabilities merge, typically because they all converge to 1 on the true hypothesis. Here are two passages that have given currency to this folklore. The first comes from the now classic review article "Bayesian Statistical Inference for Psychological Research" by Edwards, Lindman, and Savage:

> Although your initial opinion about future behavior of a coin may differ radically from your neighbor's, your opinion and his will ordinarily be transformed by application of Bayes' theorem to the results of a long sequence of experimental flips as to become nearly indistinguishable. (1963, p. 197)

A similar sentiment has been expressed by Suppes:

It is of fundamental importance to any deep appreciation of the Bayesian viewpoint to realize the particular form of the prior distribution expressing beliefs held before the experiment is conducted is not a crucial matter.... For the Bayesian, concerned as he is to deal with the real world of ordinary and scientific experience, the existence of a systematic method for reaching agreement is important.... The well-designed experiment is one that will swamp divergent prior distributions with the clarity and sharpness of its results, and thereby render insignificant the diversity of prior opinion. (1966, p. 204)

I take it that if this folklore were correct, the explanation of objectivity would have a justificatory resonance. The consensual degrees of belief are justified because they are the inevitable results of a rational process: let the Bayesian agents start off with whatever initial degrees of belief they like, as long as they conform to the probability calculus and as long as they don't differ too radically (as explained below), and let them update their opinions via the rule of conditionalization; as a result, they will all be driven by the accumulating evidence to the same final degrees of belief.

The folklore is based on more than pious hope and promissory notes. There are in fact hard mathematical results on merger of opinion that can be proved within the framework of a moderately tempered Bayesian personalism, characterized by the following principles:

P1 Degrees of belief satisfy the axioms of probability.

P2 Learning from experience is modeled as change of probability via strict conditionalization.

P3 All the agents of concern begin as *equally dogmatic* in that they initially assign 0's to the same elements of the probability space.

Principle (P3) can be motivated by a rule of mutual respect that enjoins members of a scientific community to accord a nonzero prior to any hypothesis seriously proposed by a member of the community.[7] Alternatively, it could be held that decisions on zero priors help to define scientific communities and that an account of scientific inference must be relativized to a community.[8]

The sort of result that Edwards, Lindman, Savage, and Suppes had in mind can be illustrated by an example adapted from Savage's *Foundations of Statistics* (1954). In this example it suffices to explicate (P1) in terms of (A1) to (A3) from chapter 1; countable additivity for Pr plays no role here, although it does figure essentially in the more sophisticated results discussed below in sections 4 and 5. Consider a coin-flipping experiment, and suppose that all of the Bayesian agents of concern accept the posit K of independently and identically distributed (IID) trials. Suppose further, in concert with (P3), that they all assign nonzero priors to the hypotheses $\{H_i\}$, where H_i states that the objective probability of heads is p_i ($p_i \neq p_j$ if $i \neq j$). And finally, assume that in conformity with Lewis's principal principle they all evaluate the likelihoods as

$$\Pr\left(\underset{j \leqslant n}{\&}\ E_j/H_i\ \&\ K\right) = p_i^m(1 - p_i)^{n-m},$$

where E_j reports the result of the jth flip and m is the total number of heads in the first n flips. Choose any one of the agents in the Bayesian community, and apply Bayes's theorem to conclude that for her the ratio of the posterior probabilities of two of the competing hypotheses is

$$\frac{\Pr(H_i/\&_{j \leqslant n} E_j\ \&\ K)}{\Pr(H_k/\&_{j \leqslant n} E_j\ \&\ K)} = \frac{p_i^m(1 - p_i)^{n-m}\ \Pr(H_i/K)}{p_k^m(1 - p_k)^{n-m}\ \Pr(H_k/K)}.$$

The strong form of the law of large numbers assures us that in almost every endless repetition of the experiment, the relative frequency of heads approaches the true value of the chance, say p_3, in the limit as $n \to \infty$.[9] As a consequence, the likelihood ratio for $i = 3$ and $k \neq 3$ almost surely blows up, which implies that $\Pr(H_3/\&_{j \leqslant n} E_j\ \&\ K) \to 1$ as $n \to \infty$. By (P2), the probability function $\Pr_n(\cdot)$ at stage n for an agent with starting probability $\Pr_0(\cdot) = \Pr(\cdot/K)$ is $\Pr(\cdot/\&_{j \leqslant n} E_j\ \&\ K)$. Thus as $n \to \infty$, the opinions of all of the agents regarding the H_i will almost surely merge, since each agent almost surely converges to certainty on the true hypothesis H_3.

Mary Hesse has objected that "the conditions of [Savage's type of convergence] theorem...are not valid for typical examples of scientific inference" (1975, p. 78). In particular, the crucial IID assumption certainly does not apply to the results of experiments addressed to nonstatistical hypotheses. Nor, as already noted above, is the assumption of objective likelihoods justified in these nonstatistical cases, save when HD testing is applicable.

While these objections are well taken, it is nevertheless true that more powerful convergence-to-certainty and merger-of-opinion results, none of which uses the questionable assumptions tagged by Hesse, are available in the statistics literature. Since the most elegant of these results use Doob's theory of martingales, I will briefly outline some of the leading ideas of this theory in the following section.

4   Convergence to Certainty and Merger of Opinion as a Consequence of Martingale Convergence

Consider a probability space in the mathematician's sense, that is, a triple $(\Omega, \mathscr{F}, \mathscr{P}\imath)$ consisting of a sample space Ω, a collection $\mathscr{F}$ of measurable subsets of Ω, and a countably additive function $\mathscr{P}\imath: \mathscr{F} \to [0, 1]$ such that $\mathscr{P}\imath(\Omega) = 1$. Let $\{X_n\}$ be a sequence of random variables (rv's) and let $\{\mathscr{F}_n\}$ be a sequence of σ fields such that $\mathscr{F}_n \subseteq \mathscr{F}_{n+1} \subseteq \mathscr{F}$.[10] The set $\{X_n\}$ is said to be a *martingale* with respect to $\{\mathscr{F}_n\}$ just in case for every n, $E(|X_n|) < \infty$, X_n is measurable with respect to $\mathscr{F}_n$, and $E(X_{n+1}/\mathscr{F}_n) = X_n$.[11] Doob's basic *martingale convergence theorem* states that for such a martingale, if $\sup_n E(|X_n|) < \infty$, then $\lim_{n \to \infty} X_n$ is finite and exists almost everywhere (a.e.) (see Doob 1971).

Doob's application of this result is simple but ingenious. Let X be an rv such that $E(|X|) < \infty$. Then the $X_n \equiv E(X/\mathscr{F}_n)$ form a martingale ("Doob's martingale") with respect to the $\mathscr{F}_n$. If we think of the $\mathscr{F}_n$ as corresponding to the information gathered up to and including stage n, then successive conditional expectations of X as we come to know more and more yield a martingale. If the particular Doob martingale satisfies $\sup_n E(|X_n|) < \infty$, the convergence theorem guarantees that $\lim_{n \to \infty} E(X/\mathscr{F}_n)$ is finite and exists a.e. Further, if $\mathscr{F}_\infty$ denotes the smallest σ field that contains all of the $\mathscr{F}_n$, then $\lim_{n \to \infty} E(X/\mathscr{F}_n) = E(X/\mathscr{F}_\infty)$ a.e. And if $\mathscr{F}_\infty = \mathscr{F}$, then $E(X/\mathscr{F}_\infty) = E(X/\mathscr{F}) = X$.

The final step was, to my knowledge, not explicitly noted by Doob himself, but probabilists took the step to be so obvious as not to require explicit mention. Take X to be the characteristic function $[H]$ corresponding to some hypothesis H, i.e., $[H](\omega) = 1$ if H is true at $\omega \in \Omega$, 0 otherwise. $E([H]) < \infty$, and $\sup_n E([H]/\mathscr{F}_n) < \infty$. So if $[H]$ is measurable and $\mathscr{F}_\infty = \mathscr{F}$, $\lim_{n \to \infty} E([H]/\mathscr{F}_n)(\omega) = [H](\omega)$ a.e. But $E([H]/\mathscr{F}_n)$ is just the conditional probability of H on the evidence gathered through stage n. So the upshot is that if the information gathered is complete enough ($\mathscr{F}_\infty = \mathscr{F}$), then almost surely the posterior probability of H will go to 1 if H is true and to 0 if H is false.

Hesse's complaints against Savage do not apply here, since IID trials and objective likelihoods have not been assumed. In effect, the washing out launders not only different estimates of priors but also different estimates of likelihoods. As with the Savage result, merger of opinion takes place because of the almost sure convergence to certainty. In both cases, however, the merger is of a very weak form. All that is guaranteed is that for almost any world, any pair of equally dogmatic Bayesian conditionalizers, any hypothesis H, and any desired $\varepsilon > 0$, there is an N such that after the agents have seen at least N pieces of data, their opinions regarding H will differ by no more than ε. Since N may depend not only on the world and on ε but also on H and on the pair of agents chosen, the merger can be far from uniform. Stronger results on merger of opinion can be derived, as will be discussed in the following section.

5 The Results of Gaifman and Snir

Gaifman and Snir (1982) have shown how to translate the results of section 4 into a setting more in harmony with the standard philosophical discussions of confirmation theory, where probabilities are assigned to sentences of some formal language and the results of experiment and observation are reported in the form of atomic sentences or truth-functional compounds of atomic sentences. Specifically, Gaifman and Snir work in a language $\mathscr{L}$ obtained by adding empirical predicates and empirical function symbols to first-order arithmetic, assumed to contain names for each of the natural numbers $\mathbb{N}$. The Gaifman and Snir *models* $\mathrm{Mod}_{\mathscr{L}}$ for $\mathscr{L}$ consist of interpretations of the quantifiers as ranging over $\mathbb{N}$, and interpretations of the k-ary empirical predicates and k-ary function symbols respectively as subsets of $\mathbb{N}^k$ and functions from $\mathbb{N}^k$ to $\mathbb{N}$. (So, for example, if 'P' is an atomic empirical predicate, Pi might be taken to assert that the ith flip in a coin flipping experiment is heads.) A sentence φ of $\mathscr{L}$ is said to be *valid* in $\mathscr{L}$ ($\models \varphi$) just in case φ is true in all $w \in \mathrm{Mod}_{\mathscr{L}}$.

We can now make our starting assumption (P1) more precise by requiring that the probability axioms (A1) to (A3) from chapter 2 hold for Gaifman and Snir's $\models$ and that countable additivity holds in the form

$$\Pr((\forall_i)\eta(i)) = \lim_{n \to \infty} \Pr\left(\underset{i \leqslant n}{\&} \eta(i) \right), \tag{A4$'$}$$

where $\eta(i)$ is an open formula whose only free variable is i. Assumption (A4$'$) is needed for the application of the martingale theorems.

For a sentence φ of $\mathscr{L}$,

$\text{mod}(\varphi) \equiv \{w \in \text{Mod}_{\mathscr{L}}: \varphi \text{ is true in } w\}$.

The family of sets $\{\text{mod}(\varphi): \varphi \text{ is a sentence of } \mathscr{L}\}$ is a field that generates a σ field $\mathscr{G}$. It is shown that for every Pr on $\mathscr{L}$ satisfying (A1) to (A4′) there is a unique countably additive $\mathscr{P}\imath$ on $\mathscr{G}$ such that $\mathscr{P}\imath(\text{mod}(\varphi)) = \text{Pr}(\varphi)$ for every φ. Then $(\text{Mod}_{\mathscr{L}}, \mathscr{G}, \mathscr{P}\imath)$ is the mathematical probability space. For a given Pr on $\mathscr{L}$, a property is said to hold a.e. just in case it holds for a set $K \subseteq \text{Mod}_{\mathscr{L}}$ such that $\mathscr{P}\imath(K) = 1$. Now for $w \in \text{Mod}_{\mathscr{L}}$ and a sentence φ, define φ^w as φ or $\neg\varphi$ according as $w \in \text{mod}(\varphi)$ or $w \in \text{mod}(\neg\varphi)$. A class of sentences Φ is said to *separate* a set $K \subseteq \text{Mod}_{\mathscr{L}}$ just in case for any two distinct $w_1, w_2 \in K$, there is a $\varphi \in \Phi$ such that $w_1 \in \text{mod}(\varphi)$ and $w_2 \in \text{mod}(\neg\varphi)$. (If $\Phi = \{\varphi_i\}, i = 1, 2, 3, \ldots$, separates $\text{Mod}_{\mathscr{L}}$ and if $\mathscr{G}_n$ are the fields generated by $\{\text{mod}(\varphi_i): i \leqslant n\}$, then $\mathscr{G}_n$ generates $\mathscr{G}$. Thus it is the separating power of the accumulating evidence that makes applicable the Doob martingale convergence results.) Finally, Pr_1 and Pr_2 are said to be *equally dogmatic* just in case $\mathscr{P}\imath_1$ and $\mathscr{P}\imath_2$ are mutually absolutely continuous (i.e., $\mathscr{P}\imath_1(A) = 0$ iff $\mathscr{P}\imath_2(A) = 0$ for any $A \in \mathscr{G}$). This implies the equal dogmatism of (P3) assumed above, but the converse does not necessarily hold unless Pr_1 and Pr_2 are definable in $\mathscr{L}$. For simplicity, then, assume that the Pr functions of the agents in the Bayesian community are definable in $\mathscr{L}$.

Using the standard martingale convergence theorems for $(\text{Mod}_{\mathscr{L}}, \mathscr{G}, \mathscr{P}\imath)$ and then transferring the results down to $\mathscr{L}$, we can establish the following result.

Gaifman and Snir Theorem Let $\Phi = \{\varphi_i\}$, $i = 1, 2, \ldots$, separate $\text{Mod}_{\mathscr{L}}$. Then for any sentence ψ of $\mathscr{L}$

1. $\text{Pr}(\psi/\&_{i \leqslant n} \varphi_i^w) \to [\psi](w)$ a.e. as $n \to \infty$,
2. if Pr′ is as equally dogmatic as Pr, then

$$\sup_\psi \left| \text{Pr}'\left(\psi \middle/ \underset{i \leqslant n}{\&} \varphi_i^w\right) - \text{Pr}\left(\psi \middle/ \underset{i \leqslant n}{\&} \varphi_i^w\right) \right| \to 0$$

a.e. as $n \to \infty$.

Call Φ in the hypothesis of the theorem an *evidence matrix.* If Φ is separating, part 1 of the theorem shows that the evidence accumulated in almost any world by successively checking the elements of the evidence matrix serves to drive the posterior probability to certainty in the limit, and this certainty is reliable in that in almost any world where the probability goes to 1 (respectively, 0), the hypothesis ψ is true (respectively, false).[12] The rate of convergence to certainty cannot be expected to be uniform over ψ. For example, take ψ_n to assert that in a countable sequence of balls drawn from a bottomless urn, all the balls up to and including the nth are red, while the rest are green. For a reasonable Pr function one would expect that as n gets larger and larger, it takes longer and longer for certainty to set in.

This makes all the more remarkable part 2 of the theorem, which says that merger of opinion between two equally dogmatic agents is uniform over ψ, or in mathematical jargon, that the distance between two equally dogmatic Pr functions, as measured in the uniform distance metric, goes to 0. Note, however, that without further restrictions one cannot hope to show that there is merger of opinion in the strong sense of uniform convergence over the set of equally dogmatic Pr functions.[13] This would be the case, for instance, if the collection of equally dogmatic Pr functions formed a closed convex set with a finite number of extremal points (see Schervish and Seidenfeld 1990). But such additional assumptions markedly reduce the scope of the explanation of objectivity.

These results do not rest on those presuppositions of Savage's type of result, which, though plausible for the coin flipping case, are highly implausible when applied to the testing of nonstatistical hypotheses. Also the distinguishability hypothesis of the theorem is satisfied if the empirical predicates and function symbols of $\mathscr{L}$ all stand for observable properties and functions and if the evidence matrix consists of a complete enumeration of the atomic observation sentences.[14] In this case the successive checking by direct observation of the elements of the evidence matrix serves to drive the convergence to certainty and the merger of opinion.

The theorem is undeniably impressive. Indeed, it seems almost too good to be true, especially when one reflects on the fact that ψ may have a quantifier structure as complicated as you like. In chapter 9 we will learn that there is a sense in which it is too good to be true.[15]

6   Evaluation of the Convergence-to-Certainty and Merger-of-Opinion Results

Some of the prima facie impressiveness of these results disappears in the light of their narcissistic character, i.e., the fact that the notion of 'almost surely' is judged by Pr. Sentence ψ may be true in the actual world w_a and in some intuitively natural neighborhood of worlds near w_a. But if $Pr(\psi) = 0$, $Pr(\psi/\&_{i \leqslant n} \varphi_i^w)$ is also 0 in all these worlds. This does not contradict the theorem, since these worlds form a set of measure 0, as judged by Pr. From the point of view of an omniscient observer, the self-congratulatory success of the Bayesian method is hollow if the zeros of the prior distribution are incorrectly assigned. The personalist will no doubt respond by noting that in real life there are no omniscient observers and by asserting that flesh-and-blood observers have no metastandard by which to judge the correctness of Pr. Be that as it may, 'almost surely' sometimes serves as a rug under which some unpleasant facts are swept, as we will see in chapter 9.

Another qualm concerns Gaifman and Snir's semantics for $\mathscr{L}$. In the usual semantics, the models $\widetilde{\mathrm{Mod}}_{\mathscr{L}}$ are not separated by the empirical atomic sentences, so the straightforward application of the theorem to empirical testing is lost. Perhaps, however, one should not worry about the models that lie in $\widetilde{\mathrm{Mod}}_{\mathscr{L}}$ but not in $\mathrm{Mod}_{\mathscr{L}}$, since they contain nonstandard integers and thus are in some sense "impossible worlds."

Leaving aside these qualms, the convergence-to-certainty results do ground that aspect of the objectivity of Bayesian inference concerned with the long-run match between opinion and reality; at least this is so for observational hypotheses. But the merger-of-opinion results do not serve to ground objectivity qua intersubjective agreement for opinions that fall short of certainty, and this for two different sorts of reasons. The first has to do with the limit character of the results. Keynes's lament that in the long run we are all dead has no sting in the present context if we can know in advance how long the long run is. But what is lacking in the results before us is any estimate of the rate of convergence. Nor does it seem possible to derive informative estimates in the present general setting. In Savage's type of example in section 3, results about the rate of concentration of the posterior distribution are readily derivable, since all the agents agree on the statistical model that serves to fix the form of the likelihoods. In IID experiments, for example, the concentration of the posterior, as measured by the reciprocal of the variance, can be expected to grow as $\sqrt{n}$. This happy circumstance will not obtain in general, especially when the hypotheses at issue are nonstatistical.

It is not just that different Bayesian agents will give different estimates of rates of convergence but that there may be no useful way to form the estimates. To form an estimate for a given possible world we need to know what kind of evidence is received and also what bits are received in what order. A statistical model in effect specifies the relevant evidence (e.g., the outcomes of repeatedly flipping a coin), and the assumption of independent or exchangeable trials says that the order does not matter. But in the general case, the relevant evidence can come in myriad forms, and within a form the order can matter crucially. Some sort of estimate of rate of convergence could be produced by averaging over the rates for different sequences of evidence strings. However, this requires a weighting of different sequences, and it is problematic whether in the general setting there exists a weighting function that will gain the allegiance of all the Bayesian agents.

The second reason that the formal merger results do not serve to ground objectivity derives from the observation that for some aspects of the objectivity problem not only is the long run irrelevant, so is the short run. Scientists often agree that a particular bit of evidence supports one theory better than another or that a particular theory is better supported by one experimental finding than another (e.g., the data from the perihelion of Mercury better confirm Einstein's general theory of relativity than either the red-shift data or the bending-of-light data). What happens in the long or the short run when additional pieces of evidence are added is irrelevant to the explanation of shared judgments about the evidential value of present evidence.

Finally, the theorem does not suffice to demonstrate even long-run convergence to certainty and merger of opinion for theoretical hypotheses, at least not if one form of the antirealists' argument from underdetermination is correct, for the failure of the crucial distinguishability premise corresponds to one plausible explication of the notion of underdetermination of theory by evidence. This topic will be explored in the following section.

7 Underdetermination and Antirealism

The twin goals of this section are to discuss merger-of-opinion results for theoretical hypotheses and to assess a popular argument for antirealism. I begin with a discussion of the latter argument.

The underdetermination of theory by observational evidence is widely thought to weigh in favor of a nonrealist interpretation of scientific theories. But upon first reflection, it is not easy to see how underdetermination supports *semantic antirealism*, i.e., the doctrine that theoretical terms lack referential status.[16] Nor is it obvious why underdetermination supports *epistemological antirealism*, i.e., the doctrine that observational evidence gives no good reason to believe theoretical propositions, even if their constituent terms are referential.[17] After all, observational assertions about the elsewhere are underdetermined by all possible observations that can be made here, while observational assertions about the future are underdetermined by all possible past observations. But nevertheless, we may have good reason to believe observational predictions about the elsewhere and elsewhen.[18] Is there, then, something special about theoretical propositions that allows the epistemological antirealist to take a principled stand that differs from a form of blanket skepticism?

I will explore one possible answer that can be given within the confines of Bayesian confirmation theory. Antirealists have typically been leery of Bayesianism, and seemingly with good reason, since there is nothing in the Bayesian machinery to prevent the assignment of high probabilities to theoretical propositions. If the Bayesian account of scientific inference should imply that inferences to unobserved observables stand or fall together with inferences to unobservables, then in Bayesian eyes, at least, epistemological antirealism would be reduced to general inductive skepticism.

The beginnings of an antirealist response are suggested by the merger-of-opinion results discussed above. The mere assignment of a high personal probability to a proposition, theoretical or observational, by some member of the scientific community does not constitute the good reasons for belief that we want from scientific inference. In particular, the supposed objectivity of scientific inference is missing. To explore this matter further, let me say that the degree of belief in a hypothesis H is *objectifiable* with respect to a class $\{\mathrm{Pr}\}$ of probability functions just in case for a.e. $w \in \mathrm{Mod}_{\mathscr{L}}$, there is a number r such that for every suitable evidence matrix $\Phi = \{\varphi_i\}$ and every $\mathrm{Pr} \in \{\mathrm{Pr}\}$, $\mathrm{Pr}(H/\&_{i \leqslant n} \varphi_i^w) \to r$ as $n \to \infty$. What constitutes a "suitable" Φ may be open to dispute among empiricists of different stripes, but for present purposes, let us take suitable Φ's to consist of enumerations of the atomic observation sentences of $\mathscr{L}$. Then the convergence-to-certainty results show that for any community of scientists who operate with equally dogmatic Pr functions and for any observational H, the degree of belief in H is objectifiable (for any such H, r may be taken to be 1 or 0). Whether or not the objectification sets in within the lifetime of the average scientist is something that the convergence results do not tell us. But at least in principle there is a long-run notion of objective degree of belief for observational propositions, whether or not we are around in the long run to achieve it.

For theoretical propositions the situation is altogether different. For a start, once theoretical terms are added to the language $\mathscr{L}$, the suitable evidence matrices will no longer serve to separate $\mathrm{Mod}_{\mathscr{L}}$, and consequently, the condition for the application of the convergence result fails.[19] To extend the convergence results to theoretical hypotheses, some assumption about observational distinguishability is needed. Call the incompatible theories T_1 and T_2 *weakly observationally distinguishable* (wod) for the models MOD just in case for any $w_1, w_2 \in$ MOD such that $w_1 \in \mathrm{mod}(T_1)$ and $w_2 \in \mathrm{mod}(T_2)$, there exists a (possibly quantified) observation sentence O such that $w_1 \in \mathrm{mod}(O)$ and $w_2 \in \mathrm{mod}(\neg O)$. If $\{T_i\}$ is a partition of theories that are pairwise wod for Gaifman and Snir's MOD $=$ $\mathrm{Mod}_{\mathscr{L}}$, then the degrees of belief in these theories will be objectifiable. For given any $T_j \in \{T_i\}$, $\mathrm{Pr}(T_j/\&_{i \leqslant n} \varphi_i^w) \to [T_j](w)$ a.e. for any suitable $\Phi = \{\varphi_i\}$.[20] But at this juncture the antirealist can interpose that the failure of wod is precisely what the underdetermination of theory by observation means (in at least one precise sense). Hence underdetermination does constitute an argument for epistemological antirealism by way of undermining the conditions needed to demonstrate the objectification of belief in theories.

This last move requires some comment. Consider the more usual and apparently stronger sense of observational distinguishability; namely, T_1 and T_2 are *strongly observationally distinguishable* (sod) for MOD just in case there is a (possibly quantified) observation sentence O such that for any $w_1, w_2 \in$ MOD, if $w_1 \in \mathrm{mod}(T_1)$ and $w_2 \in \mathrm{mod}(T_2)$, then $w_1 \in \mathrm{mod}(O)$ and $w_2 \in \mathrm{mod}(\neg O)$, i.e., relative to MOD, O is a consequence of T_1 and $\neg O$ is a consequence of T_2. Trivially, sod implies wod. If MOD is taken to be the usual models $\widehat{\mathrm{Mod}_{\mathscr{L}}}$ for $\mathscr{L}$, then a simple compactness argument shows that the implication goes in the opposite direction.[21] The parallel implication is not quite so obvious if MOD is taken to be Gaifman and Snir's $\mathrm{Mod}_{\mathscr{L}}$, since $\mathrm{Mod}_{\mathscr{L}}$ is not compact even if $\mathscr{L}$ contains empirical predicate symbols but no empirical function symbols (for example, there is no model in $\mathrm{Mod}_{\mathscr{L}}$ for $\{(\exists i)Pi, \neg P1, \neg P2, \ldots\}$, even though there is a model in $\mathrm{Mod}_{\mathscr{L}}$ for every finite subset). And in fact, if T_1 and T_2 are allowed to consist of the closure under logical implication of arbitrary sets of sentences, then the implication does not hold. But if we restrict attention to the case where T_1 and T_2 are sentences, which is the case at issue, then the implication does hold.[22]

This discussion raises problems for both the Bayesian who wants the merger-of-opinion results to have bite and for the would-be epistemological realist. To take the first problem first, it might seem that the convergence-to-certainty results for theoretical hypotheses are bootless. Either wod holds for pairs of $\{T_i\}$ or not. If it does not hold, then the convergence results do not apply. If it does hold, then the convergence results do apply but are useless, for wod entails that distinct pairs of the $\{T_i\}$ have incompatible observational consequences, so one can arrive at the true theory by simple eliminative induction without using the Bayesian apparatus. In fact, however, the latter horn of this dilemma is flawed, since sod does not necessarily mean that the observational consequences of the $\{T_i\}$ are finitely verifiable or falsifiable. And if finite verifiability and falsifiability fail, the convergence results do have some bite: one converges to certainty on T_{34}, say, by making more and more atomic observations and thereby converging to 0 on the (possibly multiply quantified) observation sentences that separate T_{34} from its rivals.[23] Of course, the worries about rates of convergence raised above apply here as well.

I now turn to a discussion of how the would-be epistemological realist might respond to the underdetermination argument. First, he could grant the force of the move from underdetermination to antirealism but maintain that underdetermination does not pose a serious threat, because either it is not widespread or else occurs in uninteresting varieties. Starting from a theory and tacking on theoretical epicycles that add no new observational predictions would produce an endless string of observationally indistinguishable theories, but this form of underdetermination is uninteresting, since the core theoretical content is the same in every case. Theories of gravitation that are observationally indistinguishable and that make interestingly different theoretical commitments can be created if they are permitted to remain silent about classical gravitational tests. But completeness (in the sense of yielding definite predictions) with respect to the phenomena belonging to the commonly agreed-upon explanatory domain of gravitation would seem to be a reasonable demand to impose on theories of gravitation worthy of consideration (see, for example, Will 1972). Indeed, it could be held to be a necessary condition for calling a set of axioms a theory of gravitation. Whether there are explanatorily complete and observationally indistinguishable theories of gravity that make interestingly different theoretical posits is a question that will be taken up in chapter 7.[24]

Second, the realist could deny that underdetermination does support epistemological antirealism by denying the antirealist's identification of good reasons to believe with objectified degree of belief in the Bayesian sense of merged posterior opinion. To repeat, past observations, even if they stretch infinitely far into the past, do not serve to objectify observational predictions about the future for a broad class $\{Pr\}$ of equally dogmatic probability functions.[25] But nonetheless, one might claim that past experience does give good reason to believe that the sun will rise tomorrow and that the emeralds seen in the dawn of this new light will be green. Similarly, the realist may hold that we can have good reasons to believe theoretical claims even if the degree of belief is not objectifiable in the technical sense offered above. I am sympathetic to this point of view, but it is unavailing in the present context, which seeks to discern the implications of Bayesianism for the realism versus antirealism controversy. For in its current stage of development, the Bayesian account of scientific inference contains no explication of objective good reasons other than the forced merger of subjective opinion or the apparently unworkable schemes for objectifying assignments of priors. The Bayesianized version of the realist versus antirealist debate thus grinds to a halt over the unresolved problem of objectivity.

8 Confirmability and Cognitive Meaningfulness

I suggested above that the epistemological antirealist who does not wish to be a vulgar skeptic may run afoul of Bayesianism, since quashing Bayesian inferences to unobservables threatens to quash inferences to unobserved observables. The strength of this objection is open to debate, but we need not settle the debate to recognize that the objection can be turned around to cast doubt on Bayesian inference. If Bayesians can assign nonzero priors to hypotheses about such unobservable entities as quarks, then it would seem that they can also assign nonzero priors to hypotheses about vital forces, devils, and deities. Consequently, Bayesianism faces the embarrassment of countenancing inductive arguments in favor of (or against) such hypotheses.

Perhaps the embarrassment can be faced down with a divide-and-conquer strategy.

Case 1. The hypothesis 'Jehovah exists and rules the world' (J) is so construed that it does make a difference for the probabilities of pieces of observational evidence E about, say, the amount of suffering in the world ($\Pr(E/J \ \& \ K) \neq \Pr(E/K)$). Then Bayes's theorem shows how and why (J) is confirmed (or disconfirmed) by E. So contrary to first impressions, we can properly speak of inductive arguments for the existence of God (see Swinburne 1979).

Case 2. 'Jehovah exists and rules the world' is construed so that it doesn't make a probabilistic difference for any observational evidence E($\Pr(E/J \ \& \ K) = \Pr(E/K)$). Then Bayes's theorem shows why (J) is immune to inductive considerations. In this case the embarrassment doesn't need to be explained away, since it doesn't arise.

The positivists and logical empiricists held that 'Jehovah exists and rules the world' and its like are not real hypotheses, since (in their jargon) these inscriptions are "cognitively meaningless." Initially the positivists favored verifiability/falsifiability as the identifying mark of the cognitively meaningful, but when this criterion ran into difficulties, they switched to confirmability/disconfirmability.[26] If the latter criterion is to be implemented through Bayesian personalism, then it must be conceded after all that 'Jehovah exists and rules the world' can be cognitively meaningful. To the extent that positivists and logical empiricists balk at such a conclusion, their views clash with the Bayesianism promoted here. Whether the clash is just another nail in the coffin of a dying philosophical movement or whether it is a mark against Bayesianism is a matter that I will leave to the reader to decide.

9   Alternative Explanations of Objectivity

I turn now to an examination of some of the alternative explanations (3) to (8) listed in section 1. The idea behind (3) is that a definitional ploy may succeed where honest theorem proving has failed. The notions of rationality and objectivity are relativized to a scientific community and 'community' is defined in terms of merger of opinion over the relevant range of hypotheses. This move threatens to reactivate the buzz of relativism I assumed at the beginning of the chapter to have been cured. Therefore, further discussion of this alternative will be postponed to chapter 8. Chapter 8 will also take up one form of (4), socialism in the guise of a rule for manufacturing a consensus by means of a prescription for aggregating opinions. It is not giving away too much to anticipate the conclusion that neither (3) nor (4) holds the answer to our prayers.

The remainder of this chapter will be devoted to a discussion of (5), the evolutionary solution; (6), modest but realistic solutions; (7), non-Bayesian solutions; and (8), retrenchment.

10   The Evolutionary Solution

The results of Savage and Doob discussed above have exercised a fascination not only because they entail merger of opinion but also because they reveal a link that joins Bayesian methods to truth and reliability. But because it is forged only in the infinite limit, the link revealed in the formal theorems is too weak.

A partnership between Darwin and Bayes might be thought to supply the missing link for the medium and short runs. The idea of the partnership is, first, that evolution has produced a species for which rapid merger of opinion (not necessarily to 1 or 0) takes place and, second, that the evolutionary story of this merger has the sought-after justificatory character in that our degrees of belief are reliable estimates of the actual frequencies of relevant events, since otherwise we would not have survived.

The ideas of van Fraassen (1983a) and Shimony (1988) mentioned in chapter 2 can be used to give an account of what it means for degrees of belief to be reliable estimates of frequencies, at least for simple atomic hypotheses. It is far from clear, however, what is meant by saying that my degree of belief of .75 in Einstein's GTR is a reliable estimate of a frequency. Talk about the frequency with which hypotheses of this sort have proven to be true is vague, but insofar as I understand it, the relevant frequency would seem to be 0. I can calibrate my degree of belief in Einstein's GTR with frequencies by finding an H for which my $\Pr(H)$ is naturally interpreted as an estimate of a frequency and for which I set $\Pr(H) = \Pr(GTR)$. But such calibration involves subjective judgments.

Even if the Darwin and Bayes partnership had an unproblematic statement, there would still be two obstacles to implementing it. In the first place, there is no obvious Darwinian edge to reliability of beliefs about the esoteric matters that lie at the core of modern science. Case after case could be cited from the history of science where scientists developed a strong consensus that the hidden springs of nature followed, at least approximately, the dictates of a certain theory only to become convinced at a later stage that the theory was badly flawed. In the second place, while there may be a class of propositions for which a rapid and accurate process for fixing degrees of belief was essential to survival during humankind's formative stages (e.g., 'Tiger near'), it isn't clear how far this class extends even into the realm of mundane affairs. Thus, despite the importance of weather to prosperity and even survival itself, historically, our weather forecasts have been notoriously unreliable. Perhaps we have prospered as a species not because of any general reliability of belief-fixing processes but because we are robust enough to tolerate or creative enough to maneuver around the consequences of the unreliabilities in this process.[27]

11   Modest but Realistic Solutions

The washout theorems studied above had the lofty aim of underwriting a global consensus, but because of their limit character, they proved to be incapable of explaining the consensus that exists now. This actual consensus is partial rather than sharp and spotty rather than global. Its partial and spotty nature make it at once easier and more difficult to explain—easier because there is less to explain, and more difficult because the explanation will not be uniform but will consist of disparate pieces. Here I will concentrate on explaining such comparative judgments as evidence E confirms H_1 more than it confirms H_2, or E_1 confirms H more than does E_2.

The former case seems especially difficult to deal with. Suppose, for example, that H_1 and H_2 are both hypotheticodeductively confirmed by E relative to the background knowledge K (i.e., $\{H_1, K\} \models E$ and $\{H_2, K\} \models E$). The incremental Bayesian confirmations of H_1 and H_2 are respectively

$\Pr(H_1/K)[(1/\Pr(E/K)) - 1]$

$\Pr(H_2/K)[(1/\Pr(E/K)) - 1]$,

and the absolute confirmations are respectively $\Pr(H_2/K)/\Pr(E/K)$ and $\Pr(H_2/K)/\Pr(E/K)$. Thus, on the Bayesian analysis, any judgment to the effect that E is better evidence (in either the incremental or absolute sense) for H_1 than for H_2 boils down to the judgment that $\Pr(H_1/K) > \Pr(H_2/K)$, and we are back in the middle of the swamp of the problem of priors.

The hope burns brighter when the case concerns the way in which different pieces of evidence bear on the same hypothesis or theory. Consider the three classical tests of Einstein's GTR. As noted in chapter 5, it is generally agreed by physicists that the evidence E_P of the advance of Mercury's perihelion gives more support to GTR than does the evidence E_B of the bending of light or the evidence E_R of the red shift. On the Bayesian analysis, the incremental confirmation values are $\Pr(GTR/K) \times [(1/\Pr(E_{P,B,R}/K)) - 1]$. Since the prior probability factor is the same in all three cases, the focus shifts to the prior-likelihood factors $\Pr(E_{P,B,R}/K)$. (Here we run smack into the problem of old evidence [see chapter 5], which is a thorn in the side of Bayesianism confirmation theory. I am just going to ignore it for present purposes.) Can we show that judgments about these prior likelihoods have an objective basis?

Here is one attempt. Imagine a complete enumeration $\{T_i\}$ of alternative theories of gravity, and suppose that each theory yields a definite prediction about the three classical tests.[28] By total probability,

$$\Pr(E_{P,B,R}/K) = \sum_i \Pr(E_{P,B,R}/T_i \,\&\, K) \times \Pr(T_i/K).$$

By assumption, the first factors in the sum on the right-hand side are all either 0 or 1, so the sum reduces to the sum of the priors of those theories that successfully explain the results of the test in question. Thus if it could be shown that the set of theories that succeed with respect to E_P is a proper subset of each of the sets of theories successful with respect to E_B and E_R, it would follow that, independently of judgments of the prior probabilities of the theories, E_P gives a better confirmational value than either E_B or E_R.

As mentioned in chapter 5, to first-order approximation, the most general stationary spherically symmetric line element can be written as

$$\mathrm{d}s^2 = [1 - (\alpha m/r) + (2\beta m^2/r^2)]\mathrm{d}t^2 - [1 + (2\gamma m/r)](\mathrm{d}x^2 + \mathrm{d}y^2 + \mathrm{d}z^2).$$

GTR sets the parameters α, β, and γ equal to 1. The perihelion shift depends on all three parameters, while the red shift depends only on α and the bending of light only on α and γ. Does it follow that any theory that successfully explains the perihelion shift must also explain the red shift and the bending of light? Not necessarily, for a theory can get the red shift and bending of light wrong but by compensating errors get the perihelion right. So it seems that our judgments in this case cannot be divorced from judgments about prior probabilities of theories.

Still, the Bayesian might claim a partial victory here on the grounds that he has to explain not why E_P gives better confirmational value than E_R or E_B (for in fact it may not) but only why it was thought that this was so. The long history of failures to explain the perihelion phenomenon (see Earman and Glymour 1991) coupled with the ready availability of multiple alternative explanations of the red shift perhaps explains why, around the time GTR was introduced, most physicists would have set $Pr(E_P/K) < Pr(E_R/K)$ and thus $Pr(GTR/E_P \& K) > Pr(GTR/E_P \& K)$. This explanation doesn't hold today, when many of the presently available members of the zoo of alternative theories of gravity explain the perihelion shift (see chapter 7).

12   Non-Bayesian Solutions

At present this is an empty label, since there aren't any extant non-Bayesian accounts of scientific inference that have proved to be viable across the broad range of cases. As one example of dashed hopes, I would cite Hempel's account of qualitative confirmation and Glymour's attempt to extend Hempel's ideas to the confirmation of theoretical hypotheses by means of bootstrapping relations. One might have hoped that Hempel's confirmation relations and Glymour's boostrapping relations, which are purely logicostructural relations, could provide at least part of the basis for objectivity. Alas, as we saw in chapter 3, the Bayesian apparatus is needed before any conclusions can be drawn about the bearing of these relations on the credibility of hypotheses. Other examples of dashed hopes could be cited, but enough tears have already been shed.

13   Retrenchment

If (1) through (7) of section 1 all fail, the only resort would seem to be a retrenchment to a more modest set of goals for a theory of confirmation and scientific inference, as suggested by conceiving the theory as constituting an inductive logic that parallels deductive logic. Deductive logic provides a neutral framework for evaluating deductive arguments. It is neutral in the sense that it doesn't tell us which contingent statements to accept as true. But it is not lacking in bite, since it does tell us that if we accept certain statements as true, then on pain of inconsistency we must accept certain other statements as true and reject still others as not true. On this analogy, inductive logic can be thought to provide a neutral framework for evaluating inductive arguments. It is neutral in that it doesn't tell us what degrees of belief to assign to contingent propositions. But it does have bite in that it tells us that if we assign such and such degrees of belief to such and such propositions, then on pain of inconsistency (i.e., incoherency) we must also assign specified degrees of belief to other propositions.

One might hope for a bit more than this from a theory of confirmation, although the more calls for work on our part. Consider the EUREKA! cartoon that appeared recently in the *Toronto Globe*. Why is the cartoon amusing? The part of the explanation of the humor relevant to present concerns is simply that there is in fact a sharp consensus about the outcomes of the "unnecessary experiments"—that is what makes them unnecessary. However, the basis of this consensus remains to be investigated. The worst-case result of the investigation would find a consensus built on sand, a consensus that obtains not because of merger of opinion forced by the accumulated evidence but because members of our community have given in to social pressures to conform. A better-case result would find a de facto washing out of priors. That is, the actually accumulated evidence does force a merger of opinion for the class of actual belief functions with which the members of our community have been endowed. But this class is relatively narrow, and when it is expanded with additional belief functions

Figure 6.1
EUREKA! created by Munro Ferguson, copyright 1989, distributed by Universal Press Syndicate

equally dogmatic with respect to ours, merger no longer takes place. Ever better cases are reached as this class expands until we reach the best-case result, where the consensus is very solid in that it arises for a maximal class of equally dogmatic belief functions that assign nonzero priors to the phenomena in question.

The results of such investigations will color the Bayesian interpretation of consensual degrees of belief. Where the consensus is one of the best-case types, the degrees of belief may deservedly be labeled as objective. But as we shade toward the worse-case end of the spectrum, scare quotes will need to be added to the label, and eventually the label may be withdrawn altogether. Whatever the decision, the Bayesian will insist that his apparatus is equal to drawing the relevant distinctions. And hankering after some form of objectivity beyond the ken of these distinctions is to hanker after the unobtainable.

Without taking any final stand on this issue, I want to agree partly with the Bayesians in insisting that investigations of the kind outlined above need to be carried out in detail. What I very much fear, however, is that these investigations will not reveal any strong Bayesian basis for claiming objectivity for the opinions so confidently announced at the beginning of this chapter or for the opinions implicitly endorsed in the EUREKA! cartoon. Certainly the discussion of the general problem of induction and the grue problem in particular shows that merger of opinion on hypotheses about the future cannot be forced even in a limit sense for a maximal class of equally dogmatic belief functions by any amount of evidence about the past, since these hypotheses are underdetermined by all past evidence. And I suspect that the evidence actually gathered forces the current consensuses in science only for a very circumscribed class of belief functions. Unless reasons can be found to privilege this class over others, the door is open to relativism, social constructivism, and other equally horrific isms.

14 Conclusion

In a certain mood I am all for upholding scientific common sense and for proclaiming that the presently available evidence does justify high confidence in the propositions that Velikovsky's *Worlds in Collision* is humbug, that space and time are relativistic rather than absolute, that the next emerald we examine will be green, etc. For those in a like mood, the drift of this chapter indicates that Bayesian personalism must either be supplemented or else rejected altogether as an account of scientific inference.

Some Bayesians would respond, "Scientific common sense be damned!" For them, there is no question of rejecting Bayesianism as an account of scientific inference, since (they proclaim) such an account must be couched in terms of degrees of belief and since what Bayesianism provides is rationality constraints on degrees of belief. Nor is there any question of supplementing Bayesianism, since to go beyond Bayesianism is to go beyond the "logic" of inductive inference. The supplementing principles must, therefore, be substantive in nature, and as Hume taught us, any justification for such principles must produce a regress or a vicious circle.

I trust that the reader of previous chapters will be convinced that the first part of this response is unacceptable. Bayesianism without a rule of conditionalization is hamstrung, but the attempted demonstrations of conditionalization do not succeed in showing that it is a constraint of rationality. And in chapter 9, I will argue that Bayesians cannot consistently maintain an attitude of evenhanded neutrality and at the same time prove merger-of-opinion and convergence-to-certainty results, for a Bayesianism strong enough to yield these results can be shown to embody what look suspiciously like substantive assumptions about the world. The principle at issue here is countable additivity. But even finite additivity does not enjoy an unquestioned status as a sine qua non of rationality (see Schick 1986 and the discussion of chapter 2 above).

I am enough of a non-Bayesian that I do not think that any a priori considerations block a non-Bayesian account of scientific inference. But when I survey the shortcomings of the non-Bayesian accounts that have been attempted, I despair that any such approach will work. In the grip of such despair, one might seek refuge in Goodman's circle: "An inductive inference...is justified by conformity to general rules, and a general rule by conformity to accepted inductive inferences. Predictions are justified if they conform to valid canons of induction; and the canons are valid if they accurately codify accepted inductive practice" (1983, p. 64). I do not doubt that this circle is virtuous rather than vicious. But the notion that the circle provides a resting place is an illusion. For the only uniformly accepted "general rules" or "canons" of induction are so near triviality as to make Goodman's circle so small that it cannot encompass any interesting scientific inferences. And it is unclear how to widen the circle without opening it to the full scope of rampant Bayesian personalism.

7 A Plea for Eliminative Induction

1 Teaching Dr. Watson to Do Induction

In the midst of the adventure recorded in "The Sign of Four," Sherlock Holmes chides his faithful companion, Dr. Watson: "How often have I said to you that when you have eliminated the impossible, whatever remains, *however improbable*, must be the truth?" In an age where confirmation theory is dominated by Bayesianism, talk of eliminative induction seems as quaint and unrealistic as...well, as a Sherlock Holmes story. I will try to correct this impression by arguing in section 2 that much of the bad press eliminative induction has received is unjustified, and that to succeed at the level of scientific theories, Bayesianism must incorporate elements of the eliminative view. I illustrate my claims in section 4 with a case study drawn from the history of twentieth-century gravitational theories. Section 5 draws some tentative morals for the philosophy of the methodology of science.

2 The Necessity of the Eliminative Element in Induction

Although Sherlock Holmes's rhetorical question for Dr. Watson has an odd ring when heard by Bayesian ears, the presupposition of the question can be given a respectable Bayesian gloss, namely, no matter how small the prior probability of a hypothesis, the posterior probability of the hypothesis goes to unity if all of the competing hypotheses are eliminated. This gloss fails to work if the Bayesian agent has been so unfortunate as to assign the true hypothesis a zero prior. Moreover, the ability to provide a Bayesian gloss does not mean that Bayesianism has any real explanatory power. Indeed, the eliminative inductivist will see the Bayesian apparatus merely as a tally device used to keep track of a more fundamental process. If the eliminativist is correct, Popper is turned upside down: Popper's account of scientific methodology emphasizes the corroboration of a hypothesis as arising from *unsuccessful* attempts at falsification of the hypothesis, but for the eliminativist it is *successful* attempts at falsifying *competing* hypotheses that count, and the success of inductivism piggybacks on this eliminative success.

The classic English detective story is the paradigm of eliminative induction at work. The suspects in the Colonel's murder are limited to the guests and servants at a country estate. The butler is eliminated because two unimpeachable witnesses saw him in the orangery at the time the Colonel was shot in the library. Lady Smyth-Simpson is eliminated because at the crucial moment she was dallying with the chauffeur in the Daimler. Etc. Only the guilty party, the Colonel's nephew, is left.

If $S_1, S_2, \ldots, S_N$ are the suspects, Bayes's theorem can be written

$$\Pr(S_k / \underset{i \leqslant n}{\&} E_i \ \& \ K) = \frac{\Pr(S_k/K) \times \Pr(\&_{i \leqslant n} E_i/S_k \ \& \ K)}{\sum_{j=1}^{N} \Pr(\&_{i \leqslant n} E_i/S_j \ \& \ K) \times \Pr(S_j/K)}, \tag{7.1}$$

where K is the background knowledge and $\&_{i \leqslant n} E_i$ is the sum of the evidence collected up to and including the nth stage of the investigation. As the evidence increases, more and more suspects are eliminated. If an alternative S_m is eliminated at stage n ($\Pr(S_m / \&_{i \leqslant n} E_i \ \& \ K) = 0$), it no longer appears in the sum in the denominator on the right-hand side of (7.1), and the probability that S_m had at the previous stage ($\Pr(S_m / \&_{i \leqslant n-1} E_i \ \& \ K)$) is redistributed over the remaining suspects. The Bayesian apparatus serves as a useful bookkeeping device to keep track of how the probabilities move around in the eliminative process, but in the end it does not matter, since eventually only one suspect, say S_{14}, remains, in which case, (7.1) implies that the posterior probability of S_{14} is 1. But Sherlock Holmes didn't need Bayes's theorem to tell him this.

Of course, it is hardly surprising that eliminative induction succeeds in this detective story, since the author carefully crafted the plot so as to assure the presence of all the elements needed to make it succeed. But we should not expect real-world science to be so accommodating. An important difference between fact and fiction is that in the real world of science we may have to confront hypotheses or theories that are not finitely falsifiable and therefore are not eliminable in Sherlock Holmes's sense.[1] Nevertheless, such a hypothesis or theory may be probabilistically eliminable in that accumulating evidence can drive its probability so low that it is no longer worth considering. Of course, such probabilistic elimination doesn't literally eliminate or kill the hypothesis, since it can be revived by further evidence. But then Sherlock Holmes's style of elimination doesn't literally kill a hypothesis either, since further evidence may reveal that the former evidence that, say, let the chauffeur off the hook was mistaken.

In any case, I want to emphasize that I am not under Holmes's illusion that induction can be turned into deduction from the evidence. Rather my goal here is to defend the meritorious features of eliminative induction against some wrongheaded attacks and in so doing to expose a gap between ideal Bayesianism and its application to real-world science. A prime

example of the sort of wrongheaded attack I have in mind is to be found
in one of the leading philosophy-of-science textbooks of the 1980s. In
it Ronald Giere wrote,

If the original alternatives are definite enough and few enough, and one can be sure
that they are all there, then one stands some chance of being able to eliminate all
but one. This is almost never the case when the alternatives are THEORETICAL
HYPOTHESES about some complex system. For a complex system there will be
infinitely many different possible hypotheses, only one of which is true. Rarely
would all the possible hypotheses be so neatly ordered that one could in some
way eliminate all but one. Usually, no matter how many possibilities one succeeds
in eliminating, there are still infinitely many left. It is impossible to get down to
only one. (1984, p. 170)[2]

Below I will to take issue with several aspects of this pessimistic assessment
of the prospects for eliminative induction. A major theme of this chapter
is that an important though largely neglected element of scientific progress
is the laying of the groundwork needed for eliminative induction, the
neat ordering of possible hypotheses, to use Giere's terminology. I will
argue also that when this goal is not within reach, Bayesian inductivism
and allied forms of inductivism will not work. But for the moment I
will focus more narrowly on two of Giere's claims.

The first claim is that in the case of theoretical hypotheses the eliminative inductivist is in a position analogous to that of Zeno's archer whose
arrow can never reach the target, for faced with an infinite number of
hypotheses, he can eliminate one, then two, then three, etc., but no matter
how long he labors, he will never get down to just one. Indeed, it is as if
the arrow never gets half way, or a quarter way, etc. to the target, since
however long the eliminativist labors, he will always be faced with an
infinite list. My response on behalf of the eliminativist has two parts. (1)
Elimination need not proceed in such a plodding fashion, for the alternatives may be so ordered that an infinite number can be eliminated in one
blow. (2) Even if we can never get down to a single hypothesis, progress
occurs if we succeed in eliminating finite or infinite chunks of the possibility space. This presupposes, of course, that we have some kind of measure,
or at least topology, on the space of possibilities. Whether it is to be
provided by Bayesian or by other means remains to be seen.

The second of Giere's claims that I want to take issue with is that
eliminative induction is worse off in the case of theoretical hypotheses than
in the case of observational generalizations. On the contrary, I think that
the strict form of eliminative induction cannot succeed in the latter case,

nor is it needed to make induction work in this case, whereas in the former case, a modified form of eliminativism that uses elements of Bayesianism is needed to ground objective inductive progress. The alert reader will derive these conclusions from results discussed in previous chapters, but for the less assiduous I will repeat the relevant results here and draw out the implications for the matter at hand.

Begin with an observational generalization of the form $H:(\forall i)Pa_i$, where 'P' is an observational predicate and $i = 1, 2, 3, \ldots$. If 'eliminate' means to falsify by a finite number of observations, then the alternative $\neg H:(\exists i)\neg Pa_i$ cannot be eliminated directly, nor can it be partitioned into subalternatives that can be directly eliminated. Nevertheless, straightforward instance induction succeeds in that if the prior probability $\Pr(H/K)$ of H is greater than 0 and if countable additivity holds, then the accumulation of positive instances drives the probability of H to 1: $\Pr(H/\underset{i\leqslant n}{\&}\, Pa_i \,\&\, K) \to 1$ as $n \to \infty$ (see chapters 2 and 4).

Now consider a theoretical hypothesis T and hypotheticodeductive positive instances E_i of T, i.e., $\{T,K\} \models E_i$, $i = 1, 2, 3, \ldots$. As long as $\Pr(E_{n+1}/\underset{i\leqslant n}{\&}\, E_i \,\&\, K) < 1$, the accumulation of new positive instances of T will serve to boost the posterior probability of T:

$$\Pr\left(T\,\middle/\,\underset{i\leqslant n-1}{\&}\, E_i \,\&\, K\right) < \Pr\left(T\,\middle/\,\underset{i\leqslant n}{\&}\, E_i \,\&\, K\right).$$

But there is no guarantee that the posterior probability of T will approach 1. To borrow once again an example from Keynes (1962), the use of the *Nautical Almanac* by navigators daily provides thousands of positive instances of Newton's theory of gravity, but successful navigation, no matter how often it is accomplished, does not by itself suffice to strongly probabilify Newton's theory.

In chapter 4 we saw some negative results concerning the probabilification of theoretical hypotheses by their positive instances. Suppose that there is a rival theory T' that, given K, is incompatible with T (i.e., $K \models \neg(T \,\&\, T')$) and that covers the same instances (i.e., $\{T',K\} \models E_i$ for all i). Then we have the following:

Fact 1 If $\Pr(T'/K) > 0$, then $\Pr(T/\underset{i\leqslant n}{\&}\, E_i \,\&\, K) \not\to 1$ as $n \to \infty$.

Fact 2 If $\Pr(T'/K) > \Pr(T/K)$, then for any n, $\Pr(T'/\underset{i\leqslant n}{\&}\, E_i \,\&\, K) > \Pr(T/\underset{i\leqslant n}{\&}\, E_i \,\&\, K)$. Thus the probability of T cannot be boosted above .5 by its instances E_i.

By contrast, if we stick to lowly observational generalizations that do not outrun the data, these negative results do not apply. Thus in the example given above of hypothesis $H: (\forall i)Pa_i$, there is no rival H' satisfying the suppositions of the facts above if we take E_i to be Pa_i.

Under what conditions can it be proved that the probability of the theoretical hypothesis T, if true, will go to 1 with accumulating positive evidence? In chapter 6 we investigated an answer that utilizes the powerful martingale-convergence theorems of Doob. To apply these theorems to prove convergence to certainty on the true theory T, we found that we needed to assume that T is part of a partition $\{T_j\}$, the members of which are observationally distinguishable in the sense that for any distinct T_m, $T_n \in \{T_j\}$, there is a sentence O_{mn} whose nonlogical vocabulary is purely observational such that $\{T_m, K\} \models O_{mn}$ while $\{T_n, K\} \models \neg O_{mn}$. But this means that the stage is set for a kind of eliminative induction. Of course, the relevant O_{mn}'s may involve multiple quantifications, in which case elimination may not succeed in the crude sense of falsification of alternatives by means of a finite number of verifications or falsifications of atomic observation sentences.[3] But supposing that atomic observations do suffice to drive the probability of the O_{mn} rapidly to 1 or 0, a kind of synthesis of Bayesianism and eliminative induction is achieved. Bayesian inductivism works to probabilify the true theory because of the effective elimination of alternative theories, and the effective elimination of alternative theories occurs because pure Bayesian inductivism works with respect to observational consequences of the theories. No useful formal results about rates of convergence to certainty on the observational generalizations are to be expected, but an evolutionary explanation of the fact that actual scientists do converge rapidly on their degrees of belief in the observational consequences is not out of the question (see chapter 6).

Needless to say, this sophisticated form of eliminativism cannot succeed without the help of the partition $\{T_j\}$, and such a partition is not to be had simply for the asking.[4] The active exploration of the space of possibilities and the classification of the alternatives in a manner that paves the way for eliminative induction is an aspect of scientific methodology that has been unjustly neglected by historians and philosophers of science. In section 4, I will present a case study of such an exploration, but before turning to concrete cases, I want to emphasize that in the absence of the conditions needed to make sophisticated eliminative induction work, Bayesianism offers *no* satisfying account of the probabilifications of scientific theories.

Both Keynes and Russell, two of the earliest advocates of a probabilistic epistemology, employed the following variant of Bayes's theorem, which holds in the HD case where $\{H, K\} \models E$:[5]

$$\Pr(T/E \ \& \ K) = 1 \Big/ \left[1 + \left(\frac{\Pr(\neg T/K)}{\Pr(T/K)} \right) \times \Pr(E/\neg T \ \& \ K) \right] \qquad (7.2)$$

If $\neg T$ has a nonnegligible prior, (7.2) shows that the posterior probability of T is large only if E is such that $\Pr(E/\neg T \ \& \ K)$ is small, which can be somewhat misleadingly glossed as saying that E would be unlikely to hold if T were false. This gloss is echoed in Giere's (1984) "condition 2" for a good test of T, which requires that E be such that the following holds:

Condition 2   If $\neg T \ \& \ K$, then very probably $\neg E$.[6]

I find condition 2 awkward, especially since Giere intends an objectivist interpretation of probability. For although I believe in objective propensity probabilities for quantum events, I don't see how objective propensities can be attached, say, to the outcomes of measuring the centenary advance of the perihelion of Mercury, save insofar as these propensity probabilities are 0 or 1. Thus it seems to me that in confirmation contexts what we need are not conditionals with objective probability consequents (as in condition 2) but conditional probabilities interpreted as degrees of belief.

How, then, are we to evaluate the crucial conditional probability $\Pr(E/\neg T \ \& \ K)$? The question becomes pressing when we take into account the fact that flesh-and-blood scientists are not logically omniscient. In chapter 5, two aspects of the failure of omniscience were distinguished: first, there may be a failure to recognize logical implications, and second, there may be a failure to perceive what alternative theories lie in the space of possibilities. We can acknowledge the second shortcoming by Shimony's (1970) device of the "catchall hypothesis." $\neg T$ is then represented as $T_1 \vee T_2 \vee \ldots \vee T_q \vee H_C$, where the T_i, $i = 1, 2, \ldots, q$, are the previously constructed alternatives to T and the catchall H_C says, in effect, that some as yet uninvented theory is true. Since H_C stands for *terra incognita*, the value of the crucial factor $\Pr(E/\neg T \ \& \ K)$ in Keynes's and Russell's form of Bayes's theorem is literally anybody's guess. This may be acceptable to the thoroughgoing Bayesian personalists, but it is unacceptable to anyone who wants to find a modicum of objectivity in scientific inference. The point here links directly to the material discussed in chapter 6. If H_C stands for a large chunk of possibility space—in effect, H_C is a long, possibly infinite, conjunction of unarticulated theories—then it may not be observationally distinguishable from $T_1, T_2, \ldots, T_q$, and consequently the convergence-to-certainty and merger-of-opinion theorems that are supposed to underwrite Bayesian objectivity do not apply.

A passage from Harold Jeffreys's *Theory of Probability* suggests that the *terra incognita* problem can be finessed.

Now in science one of our troubles is that the alternatives available for consideration are not always an exhaustive set. An unconsidered one may escape attention for centuries.... The unconsidered hypothesis, if it had been thought of, would either (1) have led to the [same] consequences $E_1, E_2, \ldots$ or (2) to different consequences at some stage. In the latter case the data would have been enough to dispose of it, and the fact that it was not thought of has done no harm. In the former case the considered and unconsidered alternatives would have the same consequences, and will presumably continue to have the same consequences. The unconsidered alternative becomes important only when it is explicitly stated and a type of observation can be found where it would lead to different predictions from the old one. The rise into importance of the theory of general relativity is a case in point. Even though we now know that the systems of Euclid and Newton need modification, it was still legitimate to base inferences on them until we knew what particular modification was needed. The theory of probability makes it possible to respect great men on whose shoulders we stand. (1961, p. 44)[7]

Jeffreys's point follows from some results on instance confirmation reviewed in chapter 4. Suppose, as before, that $\{T, K\} \models E_i$, $i = 1, 2, \ldots$ and that $\Pr(T/K) > 0$. Then we have the following:

Fact 3 $\Pr(E_{n+1}/\&_{i \leqslant n} E_i \, \& \, K) \to 1$ as $n \to \infty$ (Jeffreys).

Fact 4 $\Pr(\&_{n < j \leqslant n+m} E_j/\&_{i \leqslant n} E_i \, \& \, K) \to 1$ as $m, n \to \infty$ (Huzurbazar).

Jeffreys's point is that although the assumption of a nonzero prior for T is used in the proofs of facts 3 and 4, T itself is not mentioned in the conclusions. Thus, whether or not T is true, the correctness of past predictions of T justify confidence that the predictions of T will continue to be correct.

The Jeffreys and Huzurbazar results establish one precise version of what the instrumentalists have always urged, namely, that the instrumental success of theories can be detached from the theoretical superstructure. But if as noninstrumentalists we are interested in the probable truth of T, as opposed to its observational predictions, then the alternatives to T, considered and unconsidered, cannot be ignored.

In discussing the probabilification of Newton's theory of gravity, Russell makes a move similar to Jeffreys's. In Russell's illustration, K is the observation of the planetary motions prior to the discovery of Neptune, and E is the existence of Neptune at the place where Newtonian calculations showed that it should be (assuming Neptune to be the only source of perturbation of the inner planetary orbits). So the crucial factor $\Pr(E/\neg T \& K)$ is the probability that Neptune would be where it was, given that Newton's law of gravitation is false. Russell then writes, "Here we must make a proviso as to the sense in which we should use the word 'false.' It would not be right to take Einstein's theory as showing Newton's to be 'false' in the relevant sense. All quantitative scientific theories, when asserted, should be asserted with a margin of error; when this is done, Newton's theory of gravitation remains true of planetary motion" (1948, p. 410). Russell's remark is beside the point if we are concerned, as Russell professed to be, with the probabilification of Newton's *theory* as opposed to its observational consequences for planetary motions.

It may be useful at this juncture to summarize the gist of the twists and turns of the discussion of this section. I am not advocating the naive form of eliminative induction that Sherlock Holmes claimed to follow. Holmesian eliminativism is deficient in at least two ways. First, not only does science not have a Vicar on earth; it does not have a Conan Doyle to neatly list the suspects. But this fact of life cuts just as much against the Bayesianism form of inductivism, since, I am claiming, a persuasive form of probabilistic inductivism for scientific theories must be founded upon a parsing of the suspects. Second, in scientific cases the elimination may not consist of the simple one-two knockout deduction of a prediction and falsification of the prediction via direct observation. Some form of inductivism is typically needed to effect a confrontation between a theory and the raw data of observation and experiment. What I am advocating is a partnership between Holmes and Bayes. How this partnership might work is illustrated in section 4.

Furthermore, I would note that some of the (in)famous doctrines of popular macromethodologies of science can be usefully regarded as artifacts of the failure or success of eliminative induction. A theory may achieve the status of Kuhnian paradigm by default—not because it is the survivor of a systematic program of eliminative induction but because it has enjoyed some striking successes and because both strong competitors and any idea of how to generate them are lacking. In such circumstances it will be natural to try to dignify what scientists are doing by saying that "normal science" consists in "paradigm articulation." Making a virtue of necessity is an honorable tactic, but often the "necessity" here is simply a result of a lack of imagination on the part of scientists. Feyerabend's call for a proliferation of theories can be seen as a reaction to this situation, and although Feyerabend prefers to put an anarchistic or Dadaistic gloss on his call, it makes more prosaic sense as a first step in the investigation of the space of alternative theories, the charting of which is essential to eliminative induction.

3   Salmon's Retreat from Bayesian Inductivism

Although still professing to be a Bayesian, Wesley Salmon (1990) has come close to abandoning inductivism. His reasons stem in part from the problems discussed above in section 2, especially the problem of evaluating $Pr(E/\neg T \& K)$ or $Pr(E/H_C \& K)$. Salmon recommends that instead of trying to probabilify theories in absolute terms, we remain content to compare rival theories that have actually been proposed. For a comparison between two rivals T_1 and T_2 it suffices to know the ratio $Pr(T_1/E \& K)/Pr(T_2/E \& K)$. The immediate benefit is that in applications of Bayes's theorem in the form (2.2), there is a cancellation of the prior likelihood

$$Pr(E/K) = Pr(E/T_1 \& K) \times Pr(T_1/K) + Pr(E/T_2 \& K) \times Pr(T_2/K) + \cdots$$

$$+ Pr(E/H_C \& K) \times Pr(H_C/K),$$

which involves the troublesome $Pr(E/H_C \& K)$, and the evaluation of the resulting ratio requires only an evaluation of the prior ratio $Pr(T_1/K)/Pr(T_2/K)$ and the likelihood ratio $Pr(E/T_1 \& K)/Pr(E/T_2 \& K)$.[8]

As a fallen Bayesian, I am in no position to chide others for acts of apostasy. But I do want to note that there is a high price to pay for Salmon's form of apostasy. The first cost is that Salmon's restricted form of Bayesianism will no longer underwrite commonplace judgments of the differential-confirmational merits of various pieces of evidence, such as that the perihelion-advance evidence E_P gives better confirmational value with respect to GTR than the red-shift evidence E_R. For as noted in chapter 6, the incremental confirmational values are respectively

$\Pr(\text{GTR}/K)[(1/\Pr(E_P/K)) - 1],$

$\Pr(\text{GTR}/K)[(1/\Pr(E_R/K)) - 1],$

so that the judgment in question depends on a comparison of the prior-likelihood factors that Salmon wants to avoid. The second and perhaps greater cost is to strip Bayesianism of its applicability to decision making involving expected utility calculations, for the ordering of actions via such calculations can turn on the absolute probabilities of outcomes. And finally, if Bayesianism is to underwrite other commonplace scientific judgments, it must supply absolute as well as relative probabilities, since, for example, the verdict of the scientific community is that the probability of Velikovsky's *Worlds in Collision* theory is very low.[9]

Thus, unless Bayesianism is to be severely pruned back, I see no alternative to the sort of investigation of the possibilities recommended above. In some instances our grasp of the space of alternatives may seem firm. The case of the classic detective story is the paradigm example where all the players are known, but this happy circumstance is the result of artifice. In other cases, firmly entrenched background theories rather than artifice may supply the grasp. Thus, as regards the origins of the earth's moon, in our present state of knowledge of mechanics and planetary science, we take the possibilities to be accretion, capture, fission, impact and ejection, and combinations thereof. In other instances our grasp of the possibilities may seem infirm but firmable. Thus, nineteenth-century physicists concerned with the nature of light couldn't list the possible players, but there seemed to be a firm starting point for the investigation of theories of light, since it was assumed that light must be composed of either particles or waves. The remaining task of investigating the detailed possibilities was daunting but seemingly manageable, since long experience with particle mechanics suggested ways of treating light as corpuscles, and such phenomena as polarization focused the investigation of wave theorists on transverse waves. (The relevant possibility space for theories of light was radically altered by the advent of the quantum theory. A discussion of scientific revolutions will be postponed until chapter 8.) In still other instances our grasp of the alternatives may seem irremediably slippery and incomplete. Thus when Einstein presented his general theory of relativity in 1915, there was no firm starting point for the investigation of alternative theories of gravity. Indeed, it was not apparent how to set plausible bounds on the space of alternatives, much less how to parse the possibilities in a manner that would smooth the way for eliminative induction. In the following section I will study in some detail the self-conscious attempt made by physicists in the last twenty years to make eliminative induction work for gravitational theories.

4 Twentieth-Century Gravitational Theories: A Case Study

Gravitational research in this century does not conform to any of the currently available models of macromethodology, such as those of Kuhn, Lakatos, Laudan, and others. In particular, it is not happily fitted into Kuhn's dichotomy of normal versus revolutionary science. A "revolution in science" (as the *Times* of London declared in a banner headline) did follow upon the announcement of the British eclipse expeditions of 1919 that Einstein's GTR had correctly predicted the amount of bending of starlight as it passed near the sun.[10] But over the succeeding decades there has been nothing in the field that can be described as a major revolution, though a revolution is now brewing in various attempts to marry gravity and quantum mechanics. On the other hand, the developments in the field do not fit comfortably into the framework of Kuhn's puzzle-solving model of normal science, for although Einstein's GTR has played the leading role, it has never achieved the status of paradigm hegemony, and from 1916 onward competing theories of gravitation have always been available. By the early 1970s there were literally dozens of competitors—a veritable "zoo" of theories, as research workers in the field were wont to say. This zoo did not evolve because GTR was beset by anomalies, as Kuhn's view would suggest.

As noted above, a theory may become dominant either by default or by remaining standing when the Sherlock Holmeses of science have "eliminated the impossible." The explanation of the dominance achieved by GTR in the decades immediately following its introduction falls somewhere between these extremes. On his way to the GTR, Einstein considered and found wanting a number of other theories: gravity as a phenomenon in Minkowski space-time (Minkowski 1908), variable-speed-of-light theories of his own (Einstein 1912a, 1912b) and of Abraham's (1912) concoction, Nordström's (1912, 1913) conformally flat theories, a mutant nongenerally covariant theory worked out with the help of his friend Marcel Grossmann (Einstein and Grossmann 1913), and Mie's (1914) theory of everything. Although Einstein did not engage in a systematic exploration of alternative theories of gravity, he did offer a heuristic elimination in the form of arguments that were supposed to show that one is forced almost uniquely to the GTR if one walks the most natural path, starting from Newtonian theory and following the guideposts of relativity theory.[11] The first lesson of the special theory of relativity (so the story goes) is that bodies do not act upon one another at a distance but through the auspices of a field mechanism. Next, Einstein used the principle of equivalence to motivate the conclusion that the successful description of gravitational interactions cannot be carried out in the space-time of special relativity but must make use of a (pseudo-)Riemannian space-time. The metric potentials g_{ik} of the latter take the place of the Newtonian gravitational potential, and if one asks for the simplest second-order differential equation that can be constructed from the g_{ik} and that reduces to the Newtonian field equation (Poisson's equation) in the weak-field, slow-motion approximation, one arrives at the field equations of GTR. The early dominance of Einstein's theory was the result of the combination of the power of such naturality arguments, the success of GTR in resolving the long-standing anomaly in the motion of Mercury's perihelion, and the lack of equally natural and equally successful competitors in the years following 1915.[12]

But slowly at first and then with increasing volume, alternative theories began to appear. The reasons were in the main not connected with anomalies that beset GTR. Anomalies there were. Red-shift measurements stubbornly refused for decades to conform to Einstein's prediction, and the first solid confirmation of this effect did not come until 1960 with the experiment of Pound and Rebka using the Mossbauer effect.[13] The light-deflection prediction was not in much better shape, the measured results being scattered between 1/2 and 2 times the GTR value. But apart from an early theory by Leon Silberstein (1918) and a few other exceptions, theorists generally did not seek to avoid either the red-shift or the deflection-of-light prediction. Indeed, many tried to show that their pet theories of gravity could duplicate the GTR predictions for the three classical tests.

What, then, were the reasons for the proliferation of theories of gravitation? Some theories were proposed as serious competitors to GTR by physicists who were either unconvinced by Einstein's naturality arguments or underwhelmed by the empirical support for GTR. Other theories were "devised not so much as serious competitors to the general theory of relativity as foils against which predictions could be compared and contrasted as guides to further experiments" (Will 1974a, p. 33). The point here is that the design and assessment of experiments often depends on a knowledge of the predictions of alternative theories. Thus in 1919 Eddington was able to argue that the results of the light-deflection measurements reported by the British eclipse expeditions supported GTR because he assumed that only three outcomes were possible: the full GTR deflection value, the Newtonian 1/2 deflection, or no deflection.[14] We now know for sure, and Eddington should well have surmised, that this is a false trichotomy. Still other theories were products of attempts to explore the shape of the space of alternative theories and to arrive at a classification scheme that would lend itself to eliminative induction. Such was Ni's (1973) theory, which was constructed specifically to explode the hope that GTR could be distinguished from all viable metric theories by measuring values of the parameters that define a certain post-Newtonian limit (see below for details).

By the 1970s the denizens of this zoo of gravitational theories were so various as to elude any simple synopsis, and I will simply cite a few examples to give a flavor of the range of alternatives. There were Poincaré-type theories with particles acting at a distance in Minkowski space-time; Whitehead's theory utilizing a flat background metric η and a dynamic gravitational field $\mathbf{g}$ constructed from η and matter variables; Kustaanheimo's theory with a vector gravitational field on a flat space-time background; scalar-tensor theories of Brans and Dicke, Bergmann, and others postulating a scalar field that is generated by matter and nongravitational fields and that in turn serves as a source for the gravitational field; conformally flat theories of Nordström, Littlewood and Bergmann, Whitrow and Morduch, and others that postulate a flat background metric η and a scalar field ψ such that the physical metric $\mathbf{g} = \psi\eta$; stratified theories with preferred, conformally flat time slices; and on and on.[15]

Trying to coexist with this zoo had unpleasant consequences for astrophysicists, as described by Thorne and Will:

Since 1963 a number of astronomical discoveries and observations have forced astrophysicists to make general relativity a working tool in their theoretical model building: The cosmic microwave radiation, QSOs [quasi-stellar objects], pulsars, gravitational waves—models for all these are constrained by or involve relativistic gravity in a fundamental way.

We theorists, who wish to build models for these phenomena, are hamstrung: Experiment has not yet told us which relativistic theory of gravity is correct—general relativity, Brans-Dicke theory, one of Bergmann's ... multitudinous scalar-tensor theories, a theory which nobody has yet constructed, [etc.]. (1971, p. 595)

This quotation is taken from a paper entitled "Theoretical Frameworks for Testing Relativistic Gravity. I: Foundations," the first in a series of papers designed to lay the groundwork for systematic eliminative induction for relativistic theories of gravitation.[16] The way the program was conceived in 1974 was set out in a review article by Will:

Because of the lack of high-precision data favoring general relativity over any other theory, and because of the large (and growing) number of competing theories, there is a great need for a theoretical framework which be powerful enough to be used to design and assess experimental tests in detail, yet be general enough not to be biased in favor of general relativity. It should also provide a machinery for analyzing all theories of gravity which have been invented as alternatives to Einstein's one in the past 70 years, for classifying them, for elucidating their similarities and differences, and finally for comparing their predictions with the results of solar-system experiments. We would like to see experiment force us, with very few *a priori* assumptions about the nature of gravity, towards general relativity or some other theory. (1974b, p. 2)

The program as it evolved was even more ambitious than this passage makes it seem, for it was supposed to deal not only with the animals actually in cages in the zoo—the theories of gravity that had been invented over the past decades—but also with those beasts lurking in the bush—the theories that have yet to be invented and that may never be explicitly formulated. In this respect the program has a non-Bayesian dimension. The business of eliminating chunks of the possibility space can go forward without explicit articulation of theories in the subspace and certainly without playing the Bayesian game of assigning prior probabilities to theories. Such a game is dangerous to play at an early stage of the program, since at most a countable number of incompatible theories can be assigned nonzero priors. In effect, the use of the probability apparatus eliminates most theories (all but a set of measure zero in some natural measure on the space of possibilities) before any experiment is performed. Thus Sherlock Holmes would remind us yet again that when we have eliminated the impossible, whatever remains, however improbable, including theories that were given zero prior probabilities, must be the truth.

The initial inspiration for Thorne and Will's "theory of theories of gravitation" came from Dicke (1964, appendix 4), who assumed, among other things, that

1. space-time is a four-dimensional manifold,
2. the laws of gravitation must be stated in a generally covariant form,

3. gravity must be expressed in terms of one or more fields of tensorial character,

4. the dynamical equations of gravity must be derivable from an invariant action principle.

Thorne and Will dropped (4) and weakened (3) to a requirement implicit in (2); namely, the relevant physical quantities are geometric object fields (not necessarily tensorial in character) on the space-time manifold.[17] The resulting framework encompasses an enormous range of theories. Certainly every theory that is recognizably a theory of gravity from Newton through Einstein, all the competitors to GTR that have been invented since 1915, and an uncountable number of other theories yet to be invented can all be brought within this framework.

Despite the wide cast of its net, the resulting enterprise was nevertheless a case of what may properly be termed *local induction*. First, there was no pretense of considering all logically possible theories. Second, there was widespread if tacit agreement on the explanatory domain of a theory of gravity (i.e., on the phenomena that an adequate theory of gravity should explain). And third, there was general agreement on what auxiliary theories may be used in constructing the explanations.[18] Bayesian or Bayesian-like considerations are active in facilitating this locality, especially in helping to set the bounds on possible theories and in choosing allowable auxiliary theories. But on pain of circularity, no formal Bayesian-inductivist grounding can be given for the choice of the local frame, since, as argued in chapter 3, confirmation values depend on this choice.

Bayesian considerations also help to move the program forward, especially in the area of drawing from the raw experimental data useful conclusions with which to confront the theories of gravity. But the main business of the program, eliminative induction, is propelled by a process typically ignored in Bayesian accounts: the exploration of the possibility space, the design of classification schemes for the possible theories, the design and execution of experiments, and the theoretical analysis of what kinds of theories are and are not consistent with what experimental results.

The first step in Thorne and Will's program is to weed out all theories that do not display *basic viabilitv*, which demands four features:

Completeness. A viable theory must contain the resources to analyze all the phenomena in the explanatory domain. Thus a viable theory cannot simply postulate a value for the red shift; rather, a red shift prediction must be derivable from first principles of the theory working in conjunction with an allowable auxiliary theory of light. Milne's (1948) kinematic theory of relativity fails to pass this first hurdle.

Consistency. This means not only that the theory is free of internal self-contradictions but also that it predicts the same results when they are calculated by different methods with different allowable auxiliary hypotheses. Thus the theory must yield the same red shift result whether light is regarded as a particle or a wave. The Kustaanheimo and Nuotio (1967) theory falls at this hurdle.

Relativistic. In the limit where gravity is turned off, the nongravitational laws of a viable theory should agree with those of the STR. Of course, Newtonian and other classical theories fail to pass this hurdle.

Newtonian limit. The theory must agree with the predictions of Newton's laws of gravity in the Newtonian limit where gravitational fields are weak and the motions of gravitating bodies are slow in comparison with the velocity of light. Birkhoff's (1943) theory is ruled out, since it predicts that $v_{sound} = v_{light}$ in the Newtonian limit.

The second cut uses a combination of experiment and theoretical argument to eliminate the large chunk of the possibility space containing nonmetric theories. *Metric theories* of gravity postulate that gravity is expressed through a Lorentz-signature metric tensor **g**. More precisely, it is required that the world lines of uncharged test bodies[19] are geodesics of **g** and that in local Lorentz frames (the local freely falling frames of **g**)[20] the nongravitational laws reduce to those of STR. Metric theories are closely connected with the so-called principle of equivalence, two versions of which are relevant here. The *weak equivalence principle* (WEP) says that the world lines of test bodies in a gravitational field are independent of the composition and internal structure of the bodies. The *Einstein equivalence principle* (EEP) demands in addition that all nongravitational laws of physics are the same in every local Lorentz frame. Leonard Schiff conjectured that any theory of gravity that is complete, self-consistent, and satisfies WEP must also satisfy EEP. Since a complete and self-consistent theory is metric if and only if it is relativistic and satisfies the EEP, it follows from Schiff's conjecture that among the basically viable theories, metric theories are precisely those that fulfill WEP. No formal proof of Schiff's conjecture exists, but proofs in special cases combined with plausibility arguments make it a "fair confidence" conclusion.

If we start with this conclusion, experiments that establish WEP would serve to eliminate all nonmetric theories. The Eötvös experiment, which uses a torsion balance to compare the accelerations of two differently composed test bodies in a gravitational field, serves this purpose. Null results of extreme accuracy have been achieved by Roll, Krotov, and Dicke (1964) and by Braginski and Panov (1972). We thus arrive at the further fair-confidence conclusion that only metric theories are viable.[21]

The next round of elimination begins by classifying all metric theories in terms of predictions that can in principle be tested by solar-system experiments. For this we need an improved post-Newtonian limit that goes beyond the first-order Newtonian limit used to define basically viable theories. An initial attempt at defining a post-Newtonian limit was made by Eddington (1923), who observed that any stationary and spherically symmetric **g** field, such as presumably exists to good approximation as the exterior field of our sun, can to second order in the central mass be written as a function of three parameters α, β, γ (see also Robertson 1962 and Schiff 1967 and see chapters 5 and 6). Einstein's GTR requires that $\alpha = \beta = \gamma = 1$, while other metric theories require different values. In principle, the three classical tests—the red shift, perihelion advance, and bending of light—can be used to pin down the values of all three parameters and thus to eliminate large ranges of metric theories. A more accurate post-Newtonian limit was devised by Will (1971a, 1971b). The latest version of this "parameterized post-Newtonian" (PPN) formalism contains ten parameters (Will 1984). Solar-system experiments can be used to set limits on the values of these parameters and thus to squeeze the viable metric theories into smaller and smaller volumes of the parameter space.

In 1971 Thorne and Will contemplated the ideal limit of experimental measures of the PPN parameters that lands us at a single point of the parameter space. Three possibilities then exist:

1. There corresponds to the limit point only one viable metric theory, in which case the eliminativist program would be complete.
2. There corresponds to the limit point no viable metric theory, in which case the program would have to be rethought.
3. There corresponds to the limit point many viable metric theories.

We now know that (3) is the most likely outcome since different metric theories can have the same PPN limit. If (3) is indeed the outcome, we can contemplate carrying the eliminative program forward by devising an even more accurate post-post-Newtonian limit, etc. Here, however, a different type of fourth-round winnowing will be mentioned.

Metric theories of gravity differ as to their predictions about gravitational radiation. At least four classification schemes can be devised to use the differences to push forward the eliminativist program. The first and crudest separates theories into those that do and do not predict the existence of gravitational waves. Since most metric theories seem to fall on the do side of the cut, this scheme has little winnowing power by itself. More useful is the number of independent polarization modes allowed for gravitational waves. Einstein's GTR says two, while other metric theories say six or more. A third scheme looks at the speed of gravity waves. GTR says c, while other metric theories predict various values less than c. The fourth scheme looks at multipole moments of gravitational radiation. GTR predicts that the lowest multipole of emitted gravitational radiation is quadripole, while other metric theories imply different results.

Cosmological observations are the source of potentially powerful eliminative conclusions, since theories that agree on solar-system predictions at the PPN limit can diverge widely in the cosmological regime. Unfortunately, the auxiliary hypotheses used to interpret cosmological observations are both more numerous and more subject to doubt than those used at any previous stage of the program. To some extent this difficulty can be overcome by the fact that different theories of gravity often lead to qualitatively different predictions for cosmology. But even so, the potential of cosmological tests for winnowing the remaining theories of gravity will remain unrealized until theorists succeed in designing a classification scheme useful for comparing gravitational theories at the cosmological level and confronting them with observations; certainly no analogue of the PPN scheme for solar-system measurements presently exists for cosmology (see Will 1984, p. 414).

A schematic summary of Thorne and Will's eliminativist program is given in figure 7.1.[22]

5   Conclusion

The case study in section 4 is admittedly an extreme example in that scientists are rarely so self-conscious in the pursuit of eliminative induction as to publish in leading scientific journals papers whose only purpose is to discuss the foundations for a theory of ____ theories or to tout a framework

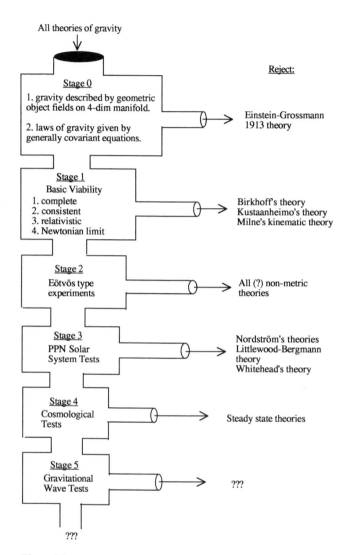

Figure 7.1
Thorne and Will's framework for testing theories of gravity

for testing theories of ____. But any number of other less self-conscious examples of eliminative induction at work could be cited. Furthermore, the general strategy can also be discerned in foundational studies. Thus, arguably little progress was made on the problem of hidden-variable interpretations of quantum mechanics until workers in the field devised various classification schemes that lend themselves to eliminativism. Hidden-variable theories can be divided into local versus nonlocal, deterministic versus stochastic, contextual versus noncontextual, etc., and a combination of mathematical results (e.g., Bell's theorems) and experiments (e.g., Aspect's experiments) can be used to rule out a chunk of the possibility space (e.g., local, noncontextual theories, whether deterministic or not). More humbly but no less importantly, having a firm grasp on whether or not to count the outcome of a measurement process as a genuine value rather than an artifact presupposes a prior grasp on both the possible sources of malfunction of the measuring instrument and the means of eliminating these sources.

The multiplication of such examples is not intended to prove, nor could it prove, the claim that either science itself or foundational studies of science typically proceed via eliminative induction, for that claim is false. A given field may not be ripe for a systematic program of eliminative induction, and I would suspect that at any given time the majority of fields, even in the advanced sciences, are unripe. What I do claim is that when the conditions for eliminative induction are not in place, there typically is no rational basis for the assignment of a high degree of confidence to some particular theory in the field. To return to the case study of section 4, my claim is that physicists of earlier decades were not rationally justified in according Einstein's GTR a high probability. And further, to be justified in this regard, it is not sufficient to get to a position where Brans and Dicke's theory, Bergmann's theories, and all of the other actually proposed alternative theories of gravity can be rejected. If a belief in GTR is to be rational, it must be based on a systematic exploration of alternatives that have yet to be invented.

The form of eliminativism I favor cuts against the variety of Bayesian personalism that focuses on how warmly one feels toward a theory and that complacently assumes that some measure of objectivity will emerge through the washing out of priors as the evidence accumulates. Only an active exploration and regimentation of the possibilities that sets the stage for eliminativism will produce the wanted objectivity. At the same time it

needs to be emphasized that my eliminativism is not identical with the
naive Sherlock Holmes variety with which I started this chapter.[23] Nor is
it anti-Bayesian per se; indeed, the numerous places where Bayesian-like
considerations come into play in the case study in section 4 indicate a
partnership that may be properly dubbed Bayesian eliminativism. However, the process described in this chapter is not compatible with the
accepted Bayesian orthodoxy. The exploration of the space of possibilities
constantly brings into consciousness heretofore unrecognized possibilities.
The resulting shifts in our belief functions cannot be described by means
of any sort of rule of conditionalization. The most dramatic shifts occur
during scientific revolutions when radically new possibilities are introduced, but the same point holds for the less dramatic, workaday cases. This
matter will be discussed in more detail in chapter 8.

In closing, I would like to offer a few remarks about the vexed topics of
scientific progress and scientific realism. As for the first, philosophers have
a penchant for accounts of the nature of scientific progress that can be
encapsulated in some pithy slogan: paradigm articulation, increase in
solved problems, progressive paradigm shift in terms of novel predictions,
etc. I believe that there is no filling for the blank in 'Scientific progress
consists in ___' that is both pithy and adequate. Scientific progress consists of all of the above and much more. Part of the more that is too often
left out of standard accounts is the conceptual advance achieved by map-
ping the topography of the space of possible theories that cover the explanatory domain in question. Whether significant dividends will accrue from
devoting more attention to this aspect of progress can only be discerned
from the fruits of historical and philosophical case studies.

Finally, I want to pick up one of the threads of the discussion from
chapter 6 on the use of underdetermination as an argument for epistemological antirealism. There I expressed my suspicion that to the extent that
the underdetermination is interesting, it is not as widespread as is commonly assumed. My suspicion can be tested by seeing what degree of
underdetermination remains when the program described in section 4 is
pushed as far as it will go. Suppose first that my suspicion proves wrong
and that a residual underdetermination of interesting scope remains after
all the solar-system and cosmological tests are exhausted. The conclusion
the eliminative inductivist would draw would be the same as that of the
epistemological antirealist: we have no good reason to believe one rather
than another of the remaining theories of gravity. The Bayesian may

demur on the grounds that the remaining theories may differ in terms of probabilities. But the only way these theories can differ in probability on the basis of evidence entailed by the theories together with the background assumptions is for the theories to differ in prior probabilities. And if the history of twentieth-century gravitational theories to the present is any guide, the theories that remain at the end of the program will be assigned significantly different priors by different scientists, which leaves the Bayesian either to identify the "correct" prior assignments or else to accept a relativism according to which, on the basis of the same evidence, Jones is justified in believing T_1 while Brown is justified in believing T_2. The first alternative is not promising, since no satisfactory account of "correct" prior assignments is in the offing. The latter alternative is even worse, since it abandons even the most minimal form of scientific objectivity.

Suppose, on the other hand, that my suspicion proves correct in that the program of section 4 converges on, say, GTR or some subset of theories that share with GTR the central theoretical mechanism of gravitation. The antirealist may respond that since the program of section 4 is a case of local induction, we have good reason to believe GTR only if the framework assumptions of the program are correct and, in particular, only if the true theory of gravity is to be found within the confines of the possibility space carved out by the framework. But since the history of science is a record of the demise of one framework after another, the program operates on unfirm ground. I have two responses to make here. First, I would contend that all cases of scientific inquiry, whether into the observable or the unobservable, are cases of local induction. Thus the present form of skepticism of the antirealist is indistinguishable from a blanket form of skepticism about scientific knowledge. This contention requires detailed argumentation, which cannot be provided here. Second, however, I do acknowledge the presuppositional character of the program. And I concede that there can be no noncircular inductive justification for circumscribing the possibility space in the fashion presupposed by the program. I am thus committed to a kind of epistemic relativism where enquiry is conducted relative to a frame for local induction, and in keeping with the discussion in section 6 of chapter 3, it seems to me that the choice of frame is best described as a pragmatic one. Before hackles rise, let me hasten to add that my relativism is the commonsensical kind entailed by any nonfoundationalist approach to knowledge that recognizes that enquiry moves forward from where we are rather than backward in search of an indubitable basis. And the pragmatic factors that guide the choice of frame can be much more weighty than mere convenience. In the case in point, the choice of the frame for investigating theories of gravitation has the full weight of the history of modern science behind it. But that weight makes itself felt not in terms of precise inductive arguments but rather in terms of Kuhnian factors, or so I will argue in the next chapter.

8 Normal Science, Scientific Revolutions, and All That: Thomas Bayes versus Thomas Kuhn

1 Kuhn's *Structure of Scientific Revolutions*

Read in the context of the then prevailing orthodoxy of logical empiricism, the first edition (1962) of Kuhn's *Structure of Scientific Revolutions* seemed to offer a novel and indeed radical account of the nature of scientific change. For those who do not have a copy of *Structure* at hand, here is a sample of a few of the purple passages:

Like the choice between competing political institutions, that between competing paradigms proves to be a choice between incompatible modes of community life.... When paradigms enter, as they must, into a debate about paradigm choice, their role is necessarily circular. Each group uses its own paradigm to argue in that paradigm's defense. (P. 94)

As in political revolutions, so in paradigm choice—there is no standard higher than the assent of the relevant community. To discover how scientific revolutions are effected, we shall therefore have to examine not only the impact of nature and logic, but also the techniques of persuasive argumentation within the quite special groups that constitute the community of scientists. (P. 94)

The proponents of competing paradigms practice their trades in different worlds.... Practicing in different worlds, the two groups of scientists see different things when they look from the same point in the same direction. (P. 150)

In these matters neither proof nor error is at issue. The transfer of allegiance from paradigm to paradigm is a conversion experience that cannot be forced. (P. 151)

Before they can hope to communicate fully, one group or the other must experience the conversion that we have been calling a paradigm shift. Just because it is a shift between incommensurables, the transition between competing paradigms cannot be made a step at a time, forced by logic and neutral experience. Like a gestalt switch, it must occur all at once (though not necessarily at an instant) or not at all. (P. 150)

Many readers saw in these passages an open invitation to arationality, if not outright irrationality. Thus Imre Lakatos took Kuhn to be saying that theory choice is a matter of "mob psychology" (1970, p. 178), while Dudley Shapere read Kuhn as saying that the decision to adopt a new paradigm "cannot be based on good reasons" (1966, p. 67). Kuhn, in turn, was equally shocked by such criticisms. In the Postscript to the second edition (1970) of *Structure* he professed surprise that readers could have imposed such unintended interpretations on the above quoted passages. Let us agree to leave aside the unfruitful question of whether or not Kuhn ought to have anticipated such interpretations and to concentrate instead on what, upon reflection, he intended to say.

Kuhn's own explanation in the Postscript begins with the commonplace that "debate over theory-choice cannot be cast in a form that resembles logical or mathematical choice" (p. 199). But he hastens to add that this commonplace does not imply that "there are no good reasons for being persuaded or that these reasons are not ultimately decisive for the group."[1] The reasons listed in the Postscript are accuracy, simplicity, and fruitfulness. The later paper "Objectivity, Value Judgments, and Theory Choice" (1977) added two further reasons: consistency and scope. And as Kuhn himself notes, the final list does not differ (with one notable exception to be discussed later) from similar lists drawn from standard philosophy-of-science texts (see also Kuhn 1983).

These soothing sentiments serve to deflate charges of arationality and irrationality, but at the same time they also serve to raise the question of how Kuhn's views are to be distinguished from the orthodoxy that *Structure* was supposed to upset. The answer given in the Postscript contains two themes elaborated in "Objectivity." First, the items on the above list are said to "function as values" that can "be differently applied, individually and collectively, by men who concur in honoring them" (p. 199). Thus, "There is no neutral algorithm for theory-choice, no systematic decision procedure which, properly applied, must lead each individual in the group to the same decision." Second, it follows (supposedly) that "it is the community of specialists rather than the individual members that makes the effective decision" (p. 200).

I think that Kuhn is correct in locating objectivity in the community of specialists, at least in the uncontroversial sense that intersubjective agreement among the relevant experts is a necessary condition for objectivity.[2] But how the community of experts reaches a decision when the individual members differ on the application of shared values is a mystery that to my mind is not adequately resolved by *Structure* or by subsequent writings. I will have more to say about this and related mysteries below.

My strategy will be to explore these and other issues raised by *Structure* from the perspective of Bayesian methodology. Before launching the exploration, I need to make some remarks about two of the most notable and controversial doctrines of *Structure*: the incommensurability of paradigms and the nonexistence of a theory-neutral observation language.

2 Theory-Ladenness of Observation and the Incommensurability of Theories

Part of what is meant by the theory-ladenness of observation is embodied in the thesis that what we see depends upon what we believe, a thesis open to challenge (see Fodor 1984). I am concerned rather with the related thesis of the nonexistence of a neutral observation language in which different theories can be compared. My response, for present purposes, is tactical. That is to say, without trying to adjudicate the general merits of the thesis, my claim is that things aren't so bad for actual historical examples. Even for cases of major scientific revolutions we can find, without having to go too far downward toward something like foundations for knowledge, an observation base that is *neutral enough for purposes at hand*. A nice example is provided by Alan Franklin (1986, pp. 110–113), who shows how to construct an experiment that is theory-neutral enough between Newtonian and special relativistic mechanics to unambiguously decide between the predictions of these theories for elastic collisions. The two theories agree on the procedure for measuring the angle between the velocity vectors of the scattered particles, and the two theories predict different angles.

More generally, I claim that in the physical sciences there is in principle always available a neutral observation base in spatial coincidences, such as dots on photographic plates, pointer positions on dials, and the like. If intersubjective agreement on such matters were not routine, then physical science as we know it would not be possible. I reject, of course, the positivistic attempt to reduce physics to such coincidences. And I readily acknowledge that such coincidences by themselves are mute witnesses in the tribunal for judging theories. But what is required to make these mute witnesses articulate is not a gestalt experience but a constellation of techniques, hypotheses, and theories: techniques of data analysis, hypotheses about the operation of measuring instruments, and auxiliary theories that support bootstrap calculations of values for the relevant theoretical parameters that test the competing theories. But I again assert that the practice is not science to the extent that this process cannot be explicitly articulated but relies on some *sui generis* form of perception. This is not to say, however, that the vulgar image of science as a blindly impartial enterprise is correct, for the articulation uncovers assumptions to which different scientists may assign very different degrees of confidence. But the sense in which different scientists can (misleadingly) be said to "see" different things when looking at the same phenomenon is one with which a probabilistic or Bayesian epistemology must cope on a routine basis, even in cases far away from the boundaries of scientific revolutions. How these differences are resolved is part of the Bayesian analogue of Kuhn's problem of community decision on theory choice. Kuhn's problem will be encountered in the following section, and the Bayesian analogue will be discussed below in sections 5 and 6.

The matter of incommensurability is much more difficult to discuss for two reasons. First, it is tied to controversial issues about meaning and reference, which I cannot broach here. Second, the topic of incommensurability presents an amorphous and shifting target. In *Structure*, for example, incommensurability was a label for the entire constellation of factors that lead proponents of different paradigms to talk past one another. In recent years Kuhn has come around to a more Carnapian or linguistic formulation in which incommensurability is equated with untranslatability. More specifically, the focus has shifted from paradigms to theories, and two theories are said to be incommensurable just in case "there is no common language into which both can be fully translated" (Kuhn 1989, p. 10). I have no doubts about Kuhn's claim that theories on different sides of a scientific revolution often use different "lexicons," that differences in lexicons can make for a kind of untranslatability, and that in turn this explains why scientists reading out-of-date texts often encounter passages that "make no sense" (1989, p. 9). But I deny that incommensurability or untranslatability in a form that makes for insuperable difficulties for confirmation or theory choice (a phrase I don't like for reasons to be given below) applies to the standardly cited cases of scientific revolutions, such as the transition from Newtonian to special-relativistic mechanics and the subsequent transition to general relativity. Newtonian, special-relativistic, general-relativistic, and theories of many other types can all be formulated in a common language, the language of differential geometry on a four-dimensional manifold, and the crucial differences in the theories lie in the differences in the geometric object fields postulated and the manner in which these fields relate to such things as particle orbits.[3] This language is anachronistic and therefore may not be the best device to use when trying to decide various historical disputes.[4] But it does seem to me to be an appropriate vehicle for framing and answering the sorts of questions of most concern to working physicists and philosophers of science. For example, on the basis of the available evidence, what is it reasonable to believe about the structure of space and time and the nature of gravitation? This is not to say that the common language makes for an easy answer. It is indeed a difficult business, but it is a business that involves the same sorts of difficulties already present when testing theories that lie on the same side of a scientific revolution. Finally, so that there can be no misunderstanding, let me repeat that I am not claiming that what I call a common language provides what Kuhn means by that term. It does not, for example, show that the Newtonian and the Einsteinian can be brought into agreement about what is and is not a "meaningful" question about simultaneity. What I do claim is that these residual elements of incommensurability do not undermine standard accounts of theory testing and confirmation.

My response to worries about the applicability of the notion of truth to whole theories is similarly local and tactical. In the Postscript to the second edition of *Structure* Kuhn writes, "There is, I think, no theory-independent way to reconstruct phrases like 'really there'; the notion of a match between the ontology of a theory and its 'real' counterpart in nature now seems to me illusive in principle" (1970, p. 206). I need not demur if 'theory' is understood in a *very* broad sense to mean something like a conceptual framework so minimal that without it "the world" would be undifferentiated Kantian ooze.[5] But I do demur if 'theory' is taken in the ordinary sense, i.e., as Newton's theory or special relativity or general relativity. For scientists are currently working in a frame in which they can say, correctly I think, that the match between the ontology of the theory and its real counterpart in nature is better for the special theory of relativity and even better for the general theory than it is for Newton's theory. Of course, to get to this position required two major conceptual revolutions. How such revolutions affect theory choice—or as I would prefer to say theory testing and confirmation—remains to be discussed.[6]

3   Tom Bayes and Tom Kuhn: Incommensurability?

Kuhn's list of criteria for theory choice is conspicuous for its omission of any reference to the degree of confirmation or probability of the theories. This is no oversight, of course, but derives both from explicit doctrines, such as the nonexistence of a theory-neutral observation language, and the largely tacit but nevertheless pervasive anti-inductivism in *Structure*. Just as striking from the Bayesian perspective is Kuhn's emphasis on theory choice or acceptance, since for the Bayesian, theories are not chosen or accepted but merely probabilified. Only in the exceptional cases where the probability is 0 or 1, or so close to one of these values as makes no odds, would there seem to be a natural commensurability between Tom Kuhn and Tom Bayes.

For the Bayesian, various problems and puzzles raised in *Structure* disappear. For example, consider the illustration Kuhn uses to reveal the difficulty of applying the criterion of theory choice most closely related to degree of confirmation: accuracy. Kuhn notes that although accuracy is the most decisive of his five criteria, it cannot uniformly discriminate among theories.

Copernicus's system ... was not more accurate than Ptolemy's until drastically revised by Kepler.... If Kepler or someone else had not found other reasons to choose heliocentric astronomy, those improvements in accuracy would never have been made, and Copernicus's work might have been forgotten. More typically, ... accuracy does permit discriminations, but not of the sort that lead regularly to unequivocal choice. The oxygen theory, for example, was universally acknowledged to account for observed weight relations in chemical reactions, something the phlogiston theory had previously scarcely attempted to do. But the phlogiston theory, unlike its rival, could account for metals' being much more alike than the ores from which they were formed. One theory thus matched experience better in one area, the other in another. (1977, p. 323)

To give a brief Bayesian commentary to this passage, we are dealing here with different times and thus with different bodies of evidence and different versions of a theory. There is no puzzle in the fact that if either of the corresponding members of the pairs (T, E) and (T', E') are different, $\Pr(T/E)$ and $\Pr(T/E')$ may be different. Nor as a Bayesian should I be worried by the fact that while on the basis of current evidence I regard the current form of T as highly probable, I regarded previous forms of T on the basis of the evidence then available as having a low (but nonzero) probability, for I was never in danger of rejecting T (or of accepting or choosing its negation), since acceptance and rejectance of theories is not my game.

A shotgun marriage of the two Toms could be arranged in either of two ways. We could take Bayes to supply the probabilities, Kuhn to supply the values or utilities, and then we could apply the rule of maximizing expected utility to render a decision on theory choice. Or we could take the Kuhnian virtues as helping to determine the probabilities—simplicity presumably affects the prior probabilities of the theories, while accuracy affects their posteriors—and then we could choose the most probable theory. But like most shotgun marriages, these would be mistakes. For Bayes, they would be mistakes because they would involve the pretense that the accepted theory T is true even though one's degree of belief in T is less than 1, perhaps substantially so. For Kuhn, they would be mistakes because the efficacy of his values in no way depends upon the truth of the theories, so estimates of the probable truth of theories are irrelevant to Kuhnian theory choice.

Part of the wrangle here derives from the unfortunate phrase 'theory choice'. Scientists do choose theories, but on behalf of the Bayesians, I would claim that they choose them only in the innocuous sense that they choose to devote their time and energy to them: to articulating them, to improving them, to drawing out their consequences, to confronting them with the results of observation and experiment.[7] Choice in this sense allows for a reconciliation of Bayes and Kuhn, since this choice is informed by both Bayesian and Kuhnian factors: probability and the values of accuracy, consistency, scope, simplicity, and fruitfulness.

Alas, this reconciliation is rather shallow. Once we are clear that the sort of choice involved in "theory choice" is a practical one, then there is nothing sacred about the list of items on Kuhn's list of values. Other values, such as getting an NSF grant or winning the Nobel Prize, can and do play a role. Further, the kind of choice in question may be bigamous, since a scientist can choose to work on two or more theories at once, and it is fickle, since it can oscillate back and forth. The kind of choice *Structure* envisioned was much more permanent; indeed, the impression given there is that normal science is not possible without tying Catholic bonds to a theory, bonds that may only be broken by leaving the Church, i.e., by creating a revolution.

Is there no way to bridge the gap between the two Toms on this issue? To explain how baffling the Bayesian finds the notion of theory acceptance, consider the case of Einstein's GTR, arguably the leading theory of gravitation and thus the top candidate for "acceptance." Marie, a research worker in the field familiar with all of the relevant experimental findings, does some introspection and finds that her degree of belief in GTR is p.

Case 1. Marie's degree of belief p is 1, or so near 1 as makes no odds. Then, as already remarked, there is a natural sense in which the Bayesian can say that Marie accepts GTR. Such cases, however, are so rare as to constitute anomalies. Of course, one can cite any number of cases from the history of science where scientists seem to be saying that for their pet theory they set $p = 1$. Here I would reissue the warning of chapter 4 that we must distinguish carefully between scientists qua advocates of theories and scientists qua judges of theories. It is the latter role that concerns us here, and in that role, scientists know, or should know, that only in very exceptional cases does the evidence rationally support a full belief in a theory.

Case 2. Marie's degree of belief p is, say, .75. Subsequently Marie decides, on the basis of her probability assignments and the values she attaches to GTR and its competitors, to "accept" GTR. What could this mean?

Subcase 2.a. When she accepts GTR, Marie changes her degree of belief from .75 to 1. This is nothing short of madness, since she has already made a considered judgment about evidential support and no new relevant evidence occasioning a rejudgment has come in.

Subcase 2.b. When she accepts GTR, Marie does not change her degree of belief from .75 to 1, but she acts *as if* all doubt were swept away in that she devotes every waking hour to showing that various puzzling astronomical observations can be explained by the theory, she assigns her graduate students research projects that presuppose the correctness of the theory, she writes a textbook on gravitational research devoted almost exclusively to GTR, etc. But at this point we have come full circle back to a sense of theory acceptance that is really a misnomer, for what is involved is a practical decision about the allocation of personal and institutional resources and not a decision about the epistemic status of the theory.[8]

This rather pedantic diatribe on theory acceptance would best be forgotten were it not for its implications for the picture of normal science. As we have seen, theory "choice" or "acceptance" can refer either to adopting an epistemic attitude or to making a practical choice. As for the former, there is no natural Bayesian explication of theory acceptance, save in the case where the probability of the theory is 1. Since scientists qua judges of theories are almost never in a position to justify such an acceptance, the Bayesian prediction is that rarely is a theory accepted in the epistemic sense. Similarly, when theory choice is a matter of deciding what theory to devote one's time and energy to, the Bayesian prediction is that in typical situations where members of the community assign different utilities to such devotions, they will make different choices. Thus, from either the epistemic or practical perspective, the Bayesian prediction is for diversity. This prediction is, I think, borne out by actual scientific practice. Thus in section 6, I will argue that insofar as normal science implies a shared paradigm, the paradigm need not, and in fact often is not, so specific as to include a particular ("accepted") theory. I will also hazard a proposal for a minimal sense of 'shared paradigm' that yields a less straitjacketed image of normal science and that also diminishes, without obliterating, the difference between normal and revolutionary science.

By way of closing this section and introducing the next, let me propose a final way of reconciling the Kuhnian and Bayesian pictures when scientific revolutions are in the offing. Radically new theories (so the story goes) carry with them different linguistic or conceptual frameworks. Thus, to even seriously entertain a new theory involves the decision to adopt, if only tentatively, the new framework. And this decision is in large part a pragmatic one, involving the factors emphasized in Kuhn's account of paradigm replacement. These considerations certainly impact on Bayesianism, since, as discussed in chapter 3, probability assignments depend on the linguistic and conceptual framework adopted. (So while it is not true, as C. I. Lewis claimed, that if anything is to be probable, then something must be certain, it is true that if anything is to be probable, something must be accepted. But that something is not a statement, whether of evidence or theory, but a framework that specifies the possibilities to be considered.) In response, let me begin by repeating my cautionary claim that major scientific revolutions need not be seen as forcing a choice between incommensurable linguistic or conceptual frameworks, since it is often possible to fit the possibilities into a larger conceptual scheme that makes the theories commensurable to the extent that there is an observation base that is neutral enough for purposes of assessing the relative confirmation of the theories. But I agree that the recognition of the larger possibility set can produce changes in the probability values and that those changes are often best described in Kuhnian terms.

4 Revolutions and New Theories

For Bayesians, a scientific revolution is not to be identified with the replacement of a paradigm in the sense of an accepted theory, since, as argued in the preceding section, Bayesians eschew theory acceptance. I suggest rather that revolutions be identified with the introductions of new theories. Such revolutions can come in one of two forms. The mildest form occurs when the new theory articulates a possibility that lay within the boundaries of the space of theories to be taken seriously but that, because of the failure of logical omniscience ((LO2) in the language of chapter 5), had previously been unrecognized as an explicit possibility. The more radical form occurs when the space of possibilities is itself significantly altered. In practice, the distinction between the two forms may be blurred, perhaps even hopelessly so, but I will begin discussion by pretending that we can separate cases.[9]

Even the mild form of revolution induces a non-Bayesian shift in belief functions. By 'non-Bayesian' I mean that no form of conditionalization, whether strict or Jeffrey or some natural extension of these, will suffice to explain the change. Conditionalizing (in any recognizable sense of the term) on the information that just now a heretofore unarticulated theory T has been introduced is literally nonsensical, for such a conditionalization presupposes that prior to this time there was a well-defined probability for this information and thus for T, which is exactly what the failure of logical omniscience rules out.

As previously noted, we can try to acknowledge the failure of logical omniscience (LO2) by means of Abner Shimony's (1970) device of a catch-all hypothesis H_C, which asserts in effect that something, we know not what, beyond the previously formulated theories $T_1, T_2, \ldots, T_q$ is true. Now suppose that a new theory T is introduced and that as a result the old degree-of-belief function Pr is changed to Pr'. The most conservative way the shift from Pr to Pr' could take place is by the process I will call *shaving off*; namely, $\Pr(T_i) = \Pr'(T_i)$ for $i = 1, 2, \ldots, q$, $\Pr'(T) = r > 0$, and $\Pr'(H_C) = \Pr(H_C) - r$. That is, under shaving off, H_C serves as a well for initial probabilities for as yet unborn theories, and the actual introduction of new theories results only in drawing upon this well without disturbing the probabilities of previously formulated theories. Unfortunately, such a conservatism eventually leads to the assignment of ever smaller initial probabilities to successive waves of new theories until a point is reached where the new theory has such a low initial probability as to stand not much of a fighting chance.

Certainly shaving off is a factually inadequate description of what happens in many scientific revolutions, especially of the more radical type. Think of what happened following the introduction of Einstein's STR in 1905. Between 1905 and 1915 little new empirical evidence in favor of STR was recorded; and yet the probability of competing theories, such as those of Lorentz and Abraham, set in classical space and time, fell in the estimates of most of the members of the European physics community, and the probability subtracted from these electron theories was transferred to Einstein's STR. The probabilities of auxiliary hypotheses may also be affected, as illustrated by the introduction of GTR. When Einstein showed that GTR accounted for the exact amount of the anomalous advance of Mercury's perihelion, the hypothesis of an amount of zodiacal matter sufficient to affect Mercury's perihelion dropped dramatically in the estimates of most of the physics community.[10]

In using the term 'non-Bayesian' to describe such nonconditionalization belief changes, whether of the conservative shaving-off type or some more radical form, I do not mean to imply that the changes are not informed by Bayesian considerations. Indeed, the problem of the transition from Pr to Pr' can be thought of as no more and no less than the familiar Bayesian problem of assigning initial probabilities, only now with a new initial situation involving a new set of possibilities and a new information basis. But the problem we are now facing is quite unlike those allegedly solved by classical principles of indifference or modern variants thereof, such as E. T. Jaynes's maximum entropy principle, where it is assumed that we know nothing or very little about the possibilities in question. In typical cases the scientific community will possess a vast store of relevant experimental and theoretical information. Using that information to inform the redistribution of probabilities over the competing theories on the occasion of the introduction of the new theory or theories is a process that is, in the strict sense of the term, *a*rational: it cannot be accomplished by some neat formal rules or, to use Kuhn's term, by an algorithm. On the other hand, the process is far from being *ir*rational, since it is informed by reasons. But the reasons, as Kuhn has emphasized, come in the form of persuasions rather than proof. In Bayesian terms, the reasons are marshalled in the guise of plausibility arguments. The deployment of plausibility arguments is an art form for which there currently exists no taxonomy. And in view of the limitless variety of such arguments, is it unlikely that anything more than a superficial taxonomy can be developed. Einstein, the consummate master of this art form, appealed to analogies, symmetry considerations, thought experiments, heuristic principles (such as the principle of equivalence), etc. All of these considerations, I am suggesting on behalf of the Bayesians, were deployed to nudge assignments of initial probabilities in favor of the theories Einstein was introducing in the early decades of this century. Einstein's success in this regard is no less important than experimental evidence in explaining the reception of his theories.

There is little to be salvaged from the Bayesian model of learning as conditionalization by claiming that, although the model fails in periods of scientific revolutions, it nevertheless holds for periods of normal science. For normal science defined as the absence of even a weak revolution shrinks to near the vanishing point. New observations, even of familiar scenes; conversations with friends; idle speculations; dreams—all of these and more are constantly introducing heretofore unarticulated possibilities and with them resultant nonconditionalization shifts in our degrees of belief, often of a non-shaving-off variety. All that remains of Bayesianism in its present form is the demand that new degrees of belief be distributed in conformity with the probability axioms. This is a nontrivial constraint, but by itself it induces only the uninteresting de Finetti form of subjectivism.

This suggests that the term 'scientific revolution' be reserved for the second and more radical form of revolution I distinguished above. For a revolution in this sense, Kuhn's purple passages do not seem overblown. The persuasions that lead to the adoption of the new shape for the possibility space cannot amount to proofs. Certainly for the Bayesian, they cannot consist of inductive proofs, since the very assignment of degrees of belief presupposes the adoption of such a space. After a revolution has taken place, the new and old theories can often be fitted into a common frame that belies any vicious form of incommensurability (as illustrated in section 2 for Newtonian and relativistic theories). But this retrospective view tends to disguise the shake-up in our system of beliefs occasioned by the adoption of the new shape for the possibility space. Bayesianism brings the shake-up to light, albeit in a way that undercuts the standard form of the doctrine.

5   Objectivity, Rationality, and the Problem of Consensus

I have endorsed a Bayesianized version of Kuhn's claim that in scientific revolutions persuasion rather than proof is the order of the day: revolutions involve the introduction of new possibilities, this introduction causes the redistribution of probabilities, the redistribution is guided by plausibility arguments, and such arguments belong to the art of persuasion.

This endorsement is confined to the first stage of the revolution, when the initial probabilities are established for the expanded possibility set. The Bayesian folklore would have it that after this first stage, something more akin to proof than persuasion operates. The idea is that an evidence-driven consensus emerges as a result of the Bayesian learning model: degrees of belief change by conditionalization on the accumulating evidence of observation and experiment, and the long-run result is that a merger of posterior opinion must take place for those Bayesian agents who initially assign zeros to the same hypotheses. In this chapter I have raised doubts about the conditionalization model. And in chapter 6, I showed why the mathematically impressive merger of opinion theorems are of dubious applicability to the sorts of cases discussed in *Structure*.

If honest theorem proving won't suffice, perhaps we can define our way to a solution. That is, why not define 'scientific community' in terms of de facto convergence of opinion over a relevant range of hypotheses? The answer is the same as that given by Kuhn in his Postscript to the threatened circularity of taking a paradigm to be what members of the community share, while also taking a scientific community to consist of just those scientists who share the paradigm. Just as scientific communities "can and should be isolated without prior recourse to paradigms" (p. 176), so they can and should be isolated without recourse to convergence-of-opinion behavior. The European physics community in the opening decades of this century can be identified by well-established historical and sociological techniques, and one wants to know how and why, for example, this community so identified reached a consensus about Einstein's STR. Nevertheless, there does seem to be at least this much truth to the definitional move: repeated failures to achieve merger of opinion on key hypotheses will most likely lead to a split in, or a disintegration of, the community.

At this juncture, let us recall Kuhn's idea that since there is "no neutral algorithm for theory-choice, no systematic procedure which, properly applied, must lead each individual in the group to the same decision ..., it is the community of specialists rather than its individual members that makes the effective decision." Even in Kuhn's own terms, I find this idea puzzling, since I do not find in *Structure* a clear account of how the group decision is to be effected. But since I have argued that there is no need to choose a theory in the choose-as-true sense and since there is no need to achieve consensus on theory choice in the choose-to-investigate-and-articulate sense, this puzzle is moot. However, the Bayesian analogue of this puzzle remains; namely, how is the community to operate so as to produce a Bayesian consensus when its members have divergent degrees of belief?

One mechanism discussed by Lehrer and Wagner in *Rational Consensus in Science and Society* (1981) requires that members of the community change their degrees of belief in accordance with a weighted-aggregation rule. Suppose that at the initial moment, person i has a degree of belief p_i^0 in the theory in question. Each person i is assumed to assign a weight $w_{ij} \geqslant 0$ to every person j, which can be taken as an index of i's opinion as to the reliability of j's opinions. According to Lehrer and Wagner's rule, i then "improves" her initial opinion p_i^0 by changing it to $p_i^1 = \sum_j w_{ij} p_j^0$. If there are still differences of opinion, the aggregation process is repeated with the p_i^1 to obtain further "improved" probabilities p_i^2, etc., until eventually the probabilities for all the members fall into line.[11]

Lehrer and Wagner offer a consistency argument for their aggregation rule: "If a person refuses to aggregate, though he does assign a positive weight to other members, he is acting as though he assigned a weight of one to himself and a weight of zero to every other member of the group. If, in fact, he assigns positive weight to other members of the group, then he should not behave as if he assigned zero weight to them" (1981, p. 22). This argument has the flavor of 'When are you going to stop beating your wife?' I do assign a positive weight to the opinions of others, but as a Bayesian I do this not by means of weighted aggregation but by conditionalization. I conditionalize on information about the opinions of my peers, and I notice that the result is a shift in my degrees of belief toward the degrees of belief of those I respect. As a young student these shifts brought my opinions closely in line with those belonging to people whom I regarded as the experts, but as a mature member of the community, I find that such shifts, while still nonnegligible, do not conform my opinions to those of others, at least not on matters where I now regard myself as an expert. And I resist any attempt to bend my carefully considered opinions.

Furthermore, there is a direct clash between Bayesianism and Lehrer and Wagner's type of rule for producing consensus. Let the 'improved' or consensus probability Pr be a weighted average $\sum_k \beta_k \mathrm{Pr}_k(\cdot)$, $0 \leqslant \beta_k \leqslant 1$ and $\sum_k \beta_k = 1$, of the individual probabilities Pr_k. Such a rule commutes with strict conditionalization only if there is dictatorship in the sense that one of the β_k's is 1 (see Berenstein, Kanal, and Lavine 1986).

Independently of Bayesianism, there are two reasons to be unhappy with the Lehrer and Wagner proposal and ones like it. The first is that it is descriptively false, as shown by the very example they use to motivate their proposal. In the 1970s Robert Dicke claimed that optical measurements of the solar disk revealed an oblateness large enough to account for $3''$ to $5''$ of arc in Mercury's centenary perihelion advance and hence to throw into doubt Einstein's explanation of the advance. When other astrophysicists disagreed with Dicke's conclusions, the differences were not smoothed over by producing a consensual probability by means of a weighted-aggregation process. The weight of opinion is now against Dicke's interpretation, but this agreement is in fact not due to aggregation but to the acquisition of additional evidence.

Of course, Lehrer and Wagner are perfectly aware of these facts, and the descriptive inadequacies of their proposal do not bother them, since they take themselves to be offering a normative proposal. But even in these terms, the proposal is to be faulted. It is fundamental to science that opinions be evidence-driven. Differences of opinion need not constitute an embarrassment that needs to be quashed, for these differences can serve as a spur to further theoretical and experimental research, and the new information produced may drive a genuine scientific consensus. If not, the attempt to manufacture a consensus by a weighted-aggregation procedure smacks of the "mob psychology" for which Kuhn was criticized.

This last point generalizes. Bayesianism, and other approaches to scientific inference as well, suggest that unless there is some evidence-driven process that operates on the level of individual scientists to produce a group consensus, the consensus will amount to something that, if not mob psychology, is nevertheless a social artifact that does not deserve either of the labels 'rational' or 'scientific'. Thus, contrary to Kuhn's idea, the group cannot decide; at least it cannot rationally decide to agree if the individuals disagree. I do not see how this conclusion can be escaped, unless some yet-to-be articulated collectivist methodology is shown to be viable.

6   A Partial Resolution of the Problem of Consensus

Part of the answer to the Bayesian version of the problem of consensus is that quite often a consensus does not exist and does not need to exist for normal scientific research to take place.[12] *Structure* warned of the danger of taking textbook science as our image of how real science actually operates, and in particular, it showed how textbook science tends to make scientific revolutions invisible by painting an overly rosy picture of a smoothly accumulating stockpile of scientific knowledge. But I think that *Structure* failed to emphasize how textbook science also disguises the diversity of opinions and approaches that flourish in nonrevolutionary science.

Consider again the case study on relativistic gravitational theory developed in chapter 7. Textbooks in this area have tended to be books on Einstein's GTR, thus fostering the illusion that GTR has achieved the status of paradigm hegemony. In addition, early textbooks not only downplayed the existence of rival theories but disguised serious difficulties with two of the principal experimental tests of GTR, the red shift and the bending of light. Normal scientific research in this field continued in the face of both a challenge (deriving from Dicke's solar-oblateness measurements) to the third experimental leg of GTR and also an ever growing number of rival theories of gravitation. This and similar examples suggest that for normal science to take place, the community of experts need only share a paradigm in the weak sense of agreement on the explanatory domain of the field, on the circumscription of the space of possible theories to be considered as serious candidates for covering the explanatory domain, on exemplars of explanatory success, and on key auxiliary hypotheses. (I am tempted to say that this is the minimal sense of paradigm needed to underwrite normal science, but historians of science probably have counterexamples waiting in the wings.)

One could argue that not having a paradigm in the stricter sense of a shared theory of gravitation has lowered the puzzle-solving efficiency of normal science. In this regard, one can recall Thorne and Will's 1971 statement that, faced with a zoo of alternative theories of gravitation, astrophysicists were hamstrung in their model-building activity. While I think that this is a fair observation, I also think that there is more to progress in normal science than puzzle solving. Chapter 7 emphasized the conceptual advances derived from the exploration of the space of possible theories, a point that brings me to the second part of the answer to the problem of consensus.

Insofar as a consensus is established, it is often due to a process akin to the eliminative induction described in chapter 7. This process is typically accompanied by a proliferation of theories not as an exercise in Feyerabendian anarchy or Dadaism but as a means of probing the possibilities and as a preliminary to developing a classification scheme that makes systematic elimination a tractable exercise. Since such elimination is not of the simpleminded Sherlock Holmes variety and involves Bayesian elements, it is well to remind ourselves of the prospects and problems of achieving a rational consensus in this way. According to the results of chapter 6, a merger of opinion for a maximal class of equally dogmatic belief functions will not be achievable, not even in the infinitely long run, if the possible theories are underdetermined by the data. Despite all of the philosophers' talk about underdetermination, its actual extent in real cases is unclear. Underdetermination aside, a consensus would be achieved by a convergence of opinion on the (possibly multiply quantified) observational predictions that separate the competing theories. Whether such a convergence takes place in the short or medium runs depends on the class of belief functions. Typically, it cannot set in rapidly for a maximal class of mutually equally dogmatic belief functions. Thus, where the consensus does obtain, one may assume that the class is less than maximal, and the more it falls below maximality, the lower the degree of rationality and objectivity the consensus will carry with it.[13]

But neither the lack of a consensus nor the less-than-solid character of the consensus where it does obtain need concern the Bayesian epistemologist. The lack of a consensus may adversely affect the social cohesion of the scientific community. But I believe that with agreement on what I termed a minimal paradigm, scientific communities are capable of much more tolerance of diversity of opinions regarding particular theories than recent philosophies of science have imagined.

7 Conclusion

The philosophy of science is littered with methodologies of science, the best known of which are associated with the names of Popper, Kuhn, Lakatos, and Laudan. In this chapter I have offered a critique of the Kuhnian version, and given the space, I would offer specific complaints about the other versions. But aside from the specifics, I have two common complaints. The first stems from the fact that each of these methodologies seizes upon one or another feature of scientific activity and tries to promote it as the centerpiece of an account of what is distinctive about the scientific enterprise. The result in each case is a picture that accurately mirrors some important facets of science but only at the expense of an overall distortion. The second common complaint is that these philosophers, as well as many of their critics, are engaged in a snark hunt in trying to find The Methodology of Science. The hunt is fueled by a conflation of three aspects of science and/or by a wrongheaded perspective on one or more of these aspects.

The first and, to my mind, the most interesting aspect is the epistemic one. I insist (in my Bayesian mode) that this aspect be explained in Bayesian terms. This implies that all valid rules of scientific inference must be derived from the probability axioms and the rule of conditionalization. It follows that there is nothing left for the methodologists to do in this area. Another implication is that the methodologists are wasting their time in searching for a demarcation criterion that will draw a bright red line between science and nonscience in terms of the methodology of belief formation and validation, for it is just all Bayesianism through and through, whether the setting is the laboratory or the street. What does demarcate science as it is now practiced is the professionalized character of its quest for well-founded belief.

This brings me to the social/institutional aspect of science, which is responsible for many of the characteristic features of scientific activity. Why, for example, do scientists display the Mertonian virtue of communalism, openly sharing information? Not because they also possess the other Mertonian virtue of disinterestedness and strive selflessly to advance scientific knowledge rather than their own agendas. On the contrary, communalism is explained by coupling the selfish desire for recognition, which obviously does motivate most scientists, with the current institutional arrangement that gives credit for a discovery to the person who first publishes it in a professional journal.[14] Such arrangements are clearly contingent, since the course of history might well have evolved a different set of protocols. And if it had evolved a very different set, science as currently practiced would not exist. Whether the practice that did evolve would deserve to be called science is a nice question that in general will not have a definite answer unless one believes, as I do not, that there are identifiable essences attached to the concept of science. I most certainly do not draw from this line of reasoning the conclusion that because they are contingent, the current social/institutional arrangements of science and the characteristics they foster are not worthy objects of study. But I do caution against trying to use the results of such a study to build an account of The Methodology of Science.

Finally, there are decisions about the tactics and strategies of scientific research, an aspect of science that the methodologists have taken as their main theater of operations. A typical issue here might (with only mild caricature) be posed thus: "My old paradigm has an impressive record of predictive and explanatory success. But lately it has been unable to generate any novel predictions that stand up to experimental test, and it has been unable to resolve several long-standing anomalies. Should I continue to tinker with it in the hope that its fortunes can be revived, or should I switch allegiance to a rival paradigm?" I suggest that this and similar issues should be seen as practical decisions about the allocation of intellectual and economic resources. From this perspective, there is nothing left for the methodologists to do except to repeat, perhaps in disguised form, the advice to choose the action that maximizes expected utility.

In sum, I agree with Feyerabend that there is no Methodology. But my reasons do not stem from an ideology of anarchism or Dadaism; nor do they rely on incommensurability and fellow travelers. A little Bayesianism and a lot of calm reflection are all that is needed.

It might be complained that the picture I have sketched leaves out the interactions among the three aspects of science I have identified. I agree that these interactions generate a number of unresolved problems. I have, for example, tried to highlight in this chapter and the preceding one the curious relationship between the epistemic and social aspects as regards the notion of scientific objectivity. A key component of scientific objectivity is agreement among members of the relevant scientific community. But an objectivity worth having requires an individualism: the consensus must emerge not from social pressures but from an evidence-driven merger of individual opinions operating under Bayesian strictures. The account I have given of the matter is far from complete, and I am unsure about what else is needed to complete the story. But I do not think that Methodology is the answer.

9 Bayesianism versus Formal-Learning Theory

1 Putnam's Diagonalization Argument

Putnam (1963a, 1963b) constructed an ingenious argument directed against Carnap's system of inductive logic. If effective, the argument would tell against Bayesianism in general. I will rehearse the argument in a form that helps to bring out these general implications.

Let me concentrate on hypotheses that can be formulated in language $\mathscr{L}$ constructed from first-order arithmetic equipped with numerals (or names for all the natural numbers) by adding names $a_1, a_2, \ldots$, intended to denote a denumerable sequence of concrete objects, and by also adding empirical predicates intended to denote properties of these objects.[1] Putnam considered two desiderata on the Pr function of any Bayesian agent who assigns degrees of belief to the propositions of $\mathscr{L}$. The first desideratum says that the Bayesian agent should be able to learn, if only in the weak sense of instance confirmation, any effective and true hypothesis.

P1 If H is an effective hypothesis and H is true, the Pr instance confirmation of H (as more and more individuals are examined) eventually becomes and remains above .5.

Some explanation is in order. Recall that instance confirmation focuses not on the probability of H itself but rather on the probability of the "next instance." The use of instance confirmation in (P1) is a concession to Carnap necessitated by the fact that in his systems of inductive logic, universal hypotheses have zero priors and thus are unlearnable in the sense that their posterior probabilities cannot go to 1. We saw in chapter 4 that this concession does not have to be made for Bayesians in general, but no harm is done in granting it here, since it only serves to strengthen the moral Putnam wants to draw. The other bit of explanation needed is that an effective hypothesis H is one such that (1) H is expressible in $\mathscr{L}$, (2) if it is a consequence of H that Ma_i (where 'M' is a molecular predicate of $\mathscr{L}$), then $H \to Ma_i$ is provable in $\mathscr{L}$, and (3) H is equivalent to a set of sentences of the form Ma_i or $\neg Ma_i$ as a_i runs through all of the individual constants.

Putnam's second desideratum is the following:

P2 For every molecular predicate 'M' of $\mathscr{L}$ and any $n \geq 0$, there is an $m > 0$ such that

$$\Pr\left(Ma_{n+m+1} \middle/ \underset{i \leqslant n}{\&} \pm Ma_i \underset{n<j \leqslant n+m}{\&} Ma_j\right) > .5.$$

That is, regardless of which among the first n individuals are M's or not, if the next m individuals are all M's for a large enough m, the probability that the next individual a_{n+m+1} is also an M is greater than .5.

In fact, (P2) follows from (P1). For if (P2) fails, there will be a hypothesis such as $H: a_1$ to a_3 are red, a_4 to a_{10} are nonred, and a_{11} and all subsequent individuals are red, such that no matter how many individuals are examined and found to conform to H, the probability of the next instance of H does not rise above .5. Since H is effective and true in some models, (P1) is violated.

Now consider an empirical predicate 'R', intended to denote (say) redness, and consider an infinite class C of integers $n_1, n_2, \ldots$ such that the Pr probability of Ra_{n_1} is greater than .5 if all the preceding individuals are red, the Pr probability of Ra_{n_2} is greater than .5 if all the preceding individuals after a_{n_1} are red, and in general the Pr probability of Ra_{n_j} is greater than .5 if all the preceding individuals after $a_{n_{j-1}}$ are red. That there is such a C follows from (P2). For if we take 'R' for 'M' and set $n = 0$, there is, by (P2), an n_1 such that if the first $(n_1 - 1)$ individuals are all red, the probability of Ra_{n_1} is greater than .5. Then again by (P2), there is an m such that if a_{n_1+1} through a_{n_1+m} are all red, the probability of Ra_{n_1+m+1} is greater than .5. Call $n_1 + m + 1$ 'n_2'. Then iterate the construction endlessly to produce an infinite class of integers. There are many such C's, but for purposes at hand, it is convenient to choose the particular C such that n_1 is the smallest number where the probability of Ra_{n_1} is greater than .5 if all the preceding individuals are red and such that n_j, for $j > 1$, is the smallest number where the probability of Ra_{n_j} is greater than .5 if all the preceding individuals after $a_{n_{j-1}}$ are red, regardless of the distribution of red and nonred among the first n_{j-1}.

Now suppose that Pr is effectively computable on truth-functional compounds of atomic empirical sentences. (I will call such a Pr *minimally recursive*.) Then C will be a recursive set, and thus by a result of Gödel, C will be the extension of a predicate 'P' explicitly definable in terms of '$+$' and '$\times$'.[2] Thus, we can formulate in $\mathscr{L}$ the diagonalization hypothesis $H_D: (\forall i)(Pi \leftrightarrow \neg Ra_i)$, where the quantification ranges over the natural numbers.[3] H_D violates (P1), for it can well be true, it is effective, but by construction its instance confirmation cannot climb and stay above .5.

Thus Carnapians in particular and Bayesians in general who use mini-mally recursive Pr functions suffer a limitation on the hypotheses they can learn, even in the weak sense of instance confirmation.

2 Taking Stock

We know from chapter 4 that the assignment of 0 probabilities to universal generalizations is an artifact of Carnap's linguistic inductive logics and not a result of the probability calculus itself. Thus it is consistent to assign $Pr(H_D) > 0$. In this case we know that a long enough unbroken string of positive instances will eventually drive the instance confirmation of H_D above .5; indeed, we know that if countable additivity is imposed, the probability of H_D itself must approach 1 in the limit as the number of positive instances approaches infinity.[4] So it must be that Putnam's assumptions force $Pr(H_D) = 0$.

Still more follows from (P1) and the assumption that Pr is minimally recursive. H_D can be divided into the conjunction of two hypotheses: H_{D1}, which says that the individuals in the infinite subsequence $a_{n_1}, a_{n_2}, \ldots$ are all nonred, and H_{D2}, which says that the individuals in the complementary subsequence are all red. If Pr treats 'R' as exchangeable for the infinite subsequence $a_{n_1}, a_{n_2}, \ldots$, then de Finetti's representation theorem says that

$$Pr(\neg Ra_{n_1} \,\&\, \neg Ra_{n_2} \,\&\, \ldots \,\&\, \neg Ra_{n_k}) = \int_0^1 \theta^k d\mu(\theta)$$

for some uniquely determined normed measure μ on $0 \leqslant \theta \leqslant 1$. Then unless μ is closed-minded in the sense that $\mu([0, \theta^*]) = 1$ for $\theta^* < .5$, the instance confirmation of H_{D1} must eventually be boosted above .5 by enough positive instances (see chapter 4). A parallel remark applies to H_{D2}. Thus Putnam's assumptions imply that Pr cannot treat 'R' as being ex-changeable over each of the two subsequences or else that the corre-sponding de Finetti measures cannot both be open-minded.

What new lesson does Putnam's construction teach us? We knew already that the power of the Bayesian apparatus derives in part from its not treating all possibilities on a par. When an uncountable number of mutual-ly exclusive possibilities is involved, to assign positive probabilities to some is perforce to assign zero probabilities to others. When the possibilities are countably infinite and when countable additivity holds, then either

some possibilities receive 0 probabilities, or else all receive positive probabilities but for any $\varepsilon > 0$ there will be an infinite number that receive a
probability less than ε. And in general, assigning positive probabilities with
seemingly desirable properties can leave some possibilities with 0 or negligible probabilities. Furthermore, to operate with exchangeability and
open-mindedness with respect to some predicates is perforce to reject exchangeability and/or open-mindedness for other predicates (see chapter 4).
What Putnam's example indicates is that there is an additional price to
be paid for the power of the Bayesian apparatus when the probability
function is required to be minimally recursive. But to fully appreciate what
the price is, it is necessary to trace out the implications of another, only
partly articulated suggestion of Putnam's paper.

3 Formal Learning Theory

Putnam's 1963a and 1963b papers were important not only for their
critique of the Carnapian and Bayesian programs but also for suggesting
a new way of thinking about inductive procedures that seemed to lie
outside the ambit of Bayesianism. This suggestion was a source for formal-
learning theory, which is now on its way to becoming a discipline in its own
right.

Let me illustrate some of the central ideas. Consider an *evidence sequence*
consisting of a countably infinite sequence of sentences drawn from that
fragment of $\mathscr{L}$ containing the sentences that will serve as the evidentiary
basis for inductive conjectures. Thus, in the setup from the preceding
sections, an example of an evidence sequence would be $\neg Ra_{11}$, $\neg Ra_2$,
$Ra_{27}, \ldots$. Such sequences are supposed to be *complete* in that they exhaust
the relevant evidence, which in terms of the present example means that
every a_i appears in the sequence. Imagine that an inductive logician is fed
such a sequence and that he is required at each stage to conjecture which
sentences of $\mathscr{L}$ (or some chosen fragment of $\mathscr{L}$) are true. Formally, then,
a *learning rule* is a function F from all initial finite segments of complete
evidence sequences to subsets of sentences of $\mathscr{L}$. If e^w stands for an evidence sequence all of whose elements are true in world w and if e_n^w stands
for the initial segment of e^w of length n, then we can say that F is *AE-
successful* (i.e., $\forall\exists$-successful) for w and for a set of sentences S just in case
for any $\psi \in S$, the truth or falsity of ψ in w is eventually identified by F, i.e.,
for any e^w and any $\psi \in S$, there is an N such that for any $m \geqslant N$, either

$F(e_m^w) \models \psi$ or $F(e_m^w) \not\models \psi$, according as ψ is true in w or ψ is false in w. F is said to be *EA-successful* for w and S just in case a time arrives at which F has identified the truth or falsity of every sentence ψ in S, i.e., for any e^w, there is an N such that for any $m \geqslant N$ and any $\psi \in S$, $F(e_m^w) \models \psi$ or $F(e_m^w) \not\models \psi$ according as ψ is true in w or ψ is false in w. F is said to be AE (respectively, EA) *reliable* with respect to S and a set K of worlds just in case it is AE (respectively, EA) successful for S and for every $w \in K$.

Many modifications and variations suggest themselves. For instance, in applications to machine learning one would want to focus on F's that are recursive or even primitive recursive. We can also contemplate demanding both more and less of a reliable learning rule. Thus, we may not be happy unless convergence to the truth takes place sufficiently rapidly. In the other direction, we may be happy if convergence is to approximate truth. And we can also contemplate that the evidence presentation is generated by some stochastic process and demand that convergence take place not invariably but with high probability.[5]

Though vague and sketchy, these remarks should suffice to indicate that Putnam's suggestions pointed the way to a rich and interesting set of problems that can be made precise and investigated by formal means. That investigation is currently under way.[6]

4 Bayesian-Learning Theory versus Formal-Learning Theory

The results discussed in chapter 6 show that Bayesians can boast of their own learning theory. Take the worlds K to be the standard models $\text{Mod}_{\mathscr{L}}$ of $\mathscr{L}$.[7] Construct an evidence sequence e^w for a world $w \in \text{Mod}_{\mathscr{L}}$ from an evidence matrix $\Phi = \{\varphi_i\}$, $i = 1, 2, \ldots$, by taking the initial segment e_n^w to be $\&_{i \leqslant n} \varphi_i^w$, where φ^w is φ or $\neg\varphi$ according as φ is true in w or φ is false in w. If the evidence matrix separates $\text{Mod}_{\mathscr{L}}$, the martingale convergence theorems yield a Bayesian-learning theorem. (Recall that Φ is said to separate $\text{Mod}_{\mathscr{L}}$ just in case for any distinct w_1, $w_2 \in \text{Mod}_{\mathscr{L}}$, there is a $\varphi_i \in \Phi$ such that φ_i is true in one of these worlds but false in the other.) For then a Bayesian agent who uses a countably additive Pr defined on $\mathscr{L}$ and who conditionalizes on the successive elements of the evidence sequence is an almost sure learner of all the true sentences of $\mathscr{L}$ in that $\Pr(\psi / \&_{i \leqslant n} \varphi_i^w) \to [\psi](w)$ for every sentence ψ and almost every $w \in \text{Mod}_{\mathscr{L}}$ as $n \to \infty$. What holds for an evidence sequence generated by w and Φ holds for the alternative sequence generated by w and a permutation Φ' of Φ, since if Φ is separating, then so is Φ'. We could make the Bayesian learner behave more like a formal learner by tacking on a rule that enjoins him to conjecture ψ just in case the probability of ψ equals or exceeds some chosen $k > .5$. But thoroughgoing Bayesians will see no need for such a device.

There remain various questions about the relation between formal-learning theory and Bayesian-learning theory that will be taken up in the remainder of this section and in the next two sections. Here I point out that there is an obvious *quasi* equivalence between the two in the context of first-order-sentence learning for complete evidence sequences. The most obvious direction is from Bayesian to formal learning. Suppose that there is a Bayesian agent Pr who, for a set of worlds $K \subseteq \text{Mod}_{\mathscr{L}}$, converges to 1 on every true sentence in a set of sentences S. Then for K and S, there is a reliable AE learning function F. For given any $w \in K$ and any $\psi \in S$ true in w, the conditional probability of ψ must eventually be driven permanently above $.5$. Thus a reliable F is given by the following rule: Conjecture ψ on any given evidence just in case the conditional probability of ψ on that evidence is greater than $.5$.

Conversely, suppose that there is a formal-learning function F that is AE-reliable for $K = \text{Mod}_{\mathscr{L}}$ and S. Then the evidence must separate K with respect to S; that is, for any $w_1, w_2 \in K$ and any $\psi \in S$, if $w_1 \in \text{mod}(\psi)$ but $w_2 \notin \text{mod}(\psi)$, then for any e^{w_1} and e^{w_2}, there must be an n such that $e_n^{w_1} \neq e_n^{w_2}$, since otherwise F would not be reliable. But given this separability, we can prove that Bayesian learning takes place for S and for some $K' \subseteq K$ of measure 1.

5   Does Formal Learning Have an Edge over Bayesian Learning?

Formal-learning theory would have an edge if there were cases where a formal learner could succeed where no Bayesian learner could. It is conceivable that the way in which the quasi equivalence outlined in the preceding section falls short of complete equivalence could be exploited to produce cases where there is a formal-learning rule that is AE-reliable for a set of sentences S and for all $\text{Mod}_{\mathscr{L}}$ whereas a Bayesian learner can reliably converge to certainty on the members of S only for a proper subset of $\text{Mod}_{\mathscr{L}}$. Putnam's example goes somewhat in this direction. Let S consist of all sentences of the form $H_n: (\forall i)[(i \geqslant n) \rightarrow Ra_i], i = 1, 2 \dots$, and also of all sentences of the form $H^Q: (\forall i)[Qi \leftrightarrow \neg Ra_i]$, where '$Q$' is a monadic predicate explicitly definable in terms of '$+$' and '$\times$' and whose extension is a recursive set of numbers. Evidence sequences are of the form $\pm Ra_1 \& \pm Ra_2 \& \ldots$ or permutations thereof. An effective AE learning function that reliably identifies the true elements of S is defined as follows. Enumerate the elements of S in any manner you like; call them $s_1, s_2, \ldots$. For any n, let $F(e_n^w)$ be the subset of those members of $\{s_1, s_2, \ldots, s_n\}$ that are consistent with e_n^w. Since at each stage n the consistency check can be made effectively, F is effective. And from the form of the elements of S and the construction of F, it is evident that for any $s_i \in S$ and any $w \in \text{Mod}_{\mathscr{L}}$, there is an N such that for any $m \geqslant N$, $F(e_m^w) \models s_i$ or not according as s_i is true in w or false in w. (Note, however, that this construction does not produce an effective and reliable EA learning function.) By contrast, there is no minimally recursive Bayesian who converges to 1 on each true H_n and H^Q for every $w \in \text{Mod}_{\mathscr{L}}$. For convergence to 1 on the true H_n guarantees that Putnam's condition (P2) will hold (i.e., regardless of which among the first n individuals are red or nonred, if the next m are all red for a large enough m, then the probability that the $(n + m + 1)$th individual will also be red is greater than .5). If the Bayesian learner is minimally recursive, we can construct his diagonalization hypothesis $H_D: (\forall i)[Pi \leftrightarrow \neg Ra_i]$. But since H_D is among the H^Q, we get a contradiction, since the probability of H_D should approach 1 in a world $w \in \text{Mod}_{\mathscr{L}}$ in which it is true, whereas by the diagonalization construction, the probability of H_D cannot be driven to stay above .5.[8]

Nevertheless, there may be a Bayesian learner, albeit a nonrecursive one, who converges to 1 on the true H_n and H^Q for every $w \in \text{Mod}_{\mathscr{L}}$. Indeed, any Bayesian whose Pr function is countably additive and assigns nonzero priors to each of the H_n and H^Q will succeed in this regard. It remains to be shown that there are such probability functions. I conjecture that they exist, but I have no proof to offer.

What holds in this example holds quite generally for first-order-sentence learning over complete evidence sequences. The existence of a formal learner who reliably AE-detects the truth values for all $\text{Mod}_{\mathscr{L}}$ of a set S of contingent sentences of $\mathscr{L}$ is matched by the existence of a reliable Bayesian learner, provided that it is possible to consistently assign nonzero priors to all members of the set S and to Boolean components of the member sentences. The proof, which will be given below, follows from a result that characterizes sentences whose truth values can be reliably identified by a formal-learning rule for all $\text{Mod}_{\mathscr{L}}$.

Sentence ψ is said to be a *verifutable* sentence just in case it is a truth-functional combination of sentences each of which has a prenex form where the quantifier prefix is either purely universal or purely existential. Evidently, there is a reliable formal-learning rule for any sentence equivalent in Mod$_{\mathscr{L}}$ to a verifutable sentence. (For a purely universal ψ, a reliable rule is this: Conjecture ψ unless and until a counterexample is found, in which case conjecture $\neg\psi$ thereafter. For a purely existential ψ, a reliable rule is this: Conjecture $\neg\psi$ unless and until a verifying instance is found, in which case conjecture ψ thereafter. For Boolean combinations of universals and existentials, do the obvious thing.)

Conversely, if ψ is not equivalent in Mod$_{\mathscr{L}}$ to a verifutable sentence, there is no formal-learning rule that reliably identifies the truth value of ψ for Mod$_{\mathscr{L}}$. For example, let 'R' now be a binary empirical predicate, and let ψ be $(\exists i)(\forall j)Ra_i a_j$. Suppose, for purposes of *reductio*, that there is a reliable F. Then, following Kelly and Glymour (1989), we proceed to diagonalize F by defining an evidence sequence e by a bait-and-switch strategy. Start with an initial finite segment e_k, and suppose for simplicity that e_k mentions individuals $a_1, \ldots, a_m$. Then if $F(e_k) \models \psi$, add to e_k a chunk that fills out in any manner you like any relations among the $a_1, \ldots, a_m$ not already fixed by e_k and that also contains the sentences $\{\neg Ra_i a_{m+1} : i \leqslant m\} \cup \{\neg Ra_{m+1} a_i : i \leqslant m\}$. If, on the other hand, $F(e_k) \not\models \psi$, add to e_k a chunk that fills out relations as before but that contains the additional sentences $\{Ra_i a_{m+1} : i \leqslant m\} \cup \{Ra_{m+1} a_i : i \leqslant m\}$. Iterate ad infinitum. Since F is supposed to identify the truth value of ψ, it must be the case that either (1) there is an N such that for all $n \geqslant N$, $F(e_n) \models \psi$ or (2) there is an N such that for all $n \geqslant N$, $F(e_n) \not\models \psi$. In case (1) it follows from the construction that in e no element is R-related to infinitely many others. Thus ψ is false, and F has failed. In case (2) at most a finite number of individuals fail to be R-related to all others. Thus ψ is true, and F has failed again.

To show that a formal learner who reliably detects that the truth value of a contingent sentence ψ for all Mod$_{\mathscr{L}}$ is matched by a reliable Bayesian learner, start from the fact that ψ must be equivalent to a vertifutable sentence. Choose a Pr function that is countably additive and assigns nonzero priors to ψ and to each Boolean component of ψ. The posterior probability for any such component will converge to the correct value, since, as we already know, the Pr value for any purely universal or purely existential sentence goes to 1 or 0 for any complete evidence sequence from

$w \in \text{Mod}_{\mathscr{L}}$ according as the sentence is true in w or false in w. Finally, to show that the Pr values work out correctly for Boolean combinations, one appeals to the facts that if $\Pr(A/E_n) \to 1$ and $\Pr(B/E_n) \to 1$ as $n \to \infty$, then $\Pr(A \ \& \ B/E_n) \to 1$ as $n \to \infty$, while if $\Pr(A/E_n) \to 0$ as $n \to \infty$, then $\Pr(A \ \& \ B/E_n) \to 0$ as $n \to \infty$.

To show that a reliable AE formal learner for a set S of contingent sentences is matched by the existence of a reliable Bayesian learner, it remains to be demonstrated that there are probability functions that consistently assign nonzero priors to all of the sentences of S and to their Boolean components. Here I have only a vague conjecture to offer: there exist such probability functions except for "perverse" S's. Of course, the label reflects my Bayesian prejudice. Whether or not it is in fact justified is a judgment that must await concrete results.

6   The Dogmatism of Bayesianism

The fact that a formal learner can reliably identify the truth of ψ for all $\text{Mod}_{\mathscr{L}}$ only if ψ is verifutable seems at first blush to condemn formal-learning theory. For science is concerned not just with such verifutable hypotheses but also with multiply quantified hypotheses, such as 'For every galaxy, there is a star with planets, and for every such planet, there is a moon such that ....'[9] It would seem then that any learning theory worthy of the name should give us a handle on such cases. In a sense, Bayesian-learning theory does, since the convergence-to-certainty theorem applies to sentences with as complex a quantificational structure as you like. But only a little reflection is needed to see that this attempted condemnation of formal-learning theory rebounds against Bayesianism.

The Bayesian convergence-to-certainty theorem does indeed apply to a sentence like $\psi \colon (\exists i)(\forall j) R a_i a_j$. And it guarantees convergence to 1 or 0 according as ψ is true in w or false in w for all w in a set $K \subseteq \text{Mod}_{\mathscr{L}}$ of measure 1. But since the existence of a Bayesian learner who is reliable for K implies the existence of a formal learner who is also reliable for K and since there is no formal learner who is reliable in this case for all of $\text{Mod}_{\mathscr{L}}$, it follows that K must be a proper subset of $\text{Mod}_{\mathscr{L}}$. The reliability we yearn for cannot be achieved even by the use of probabilistic means. And the reliability that can be achieved by probabilistic means for K can also be achieved by nonprobabilistic means.

The Bayesian might try to recoup by asserting that he has a motivated way to pick out subsets of $\text{Mod}_{\mathscr{L}}$ for which reliable learning is significant; namely, pick out subsets of measure 1 (which we may think of as the subsets of possible worlds of which the agent is probabilistically certain that they contain the actual world). The formal learner can respond that being of measure 1 is not sufficient, since there may be many such subsets on which the Bayesian is not a reliable learner. And the formal learner can conjecture that any way that the Bayesian has of characterizing the subsets of worlds on which he is reliable is matched by a nonprobabilistic characterization (e.g., $K = \text{mod}(\Delta)$ for some set Δ of sentences).

While this tit-for-tat game is inconclusive, it does leave an indelible black mark against the Bayesian side, and it is the convergence-to-certainty theorem, one of the glories of Bayesian methodology, that is the instrument that stamps the stigmata. For the theorem implies that the Bayesian is probabilistically certain that the actual world lies in a narrow enough proper subset of $\text{Mod}_{\mathscr{L}}$ that a sentence of arbitrarily complex quantificational structure can have its truth value reliably identified in the limit by a formal learning rule—something not true for $\text{Mod}_{\mathscr{L}}$ for even the simplest of nonverifutable hypotheses. An astrophysicist, for example, *may* bring to the context of inquiry enough information to guarantee that the world does lie in a narrow enough subset so as to make the truth value of 'For every galaxy, there is a star with planets, and for every such planet, there is a moon such that ...' discernible by a formal-learning rule. But it seems wrong for a method of inquiry to *require* that the astrophysicist start with such a priori knowledge.[10]

To illustrate what this a priori knowledge involves, I will derive a simple necessary condition for formal learnability relative to a $K \subseteq \text{Mod}_{\mathscr{L}}$ that is *compact in the evidence*, which means that for any set δ of evidence sentences, if for every finite subset $\delta' \subseteq \delta$, there is a $w' \in K$ such that $w' \in \text{mod}(\delta')$, then there is a $w \in K$ such that $w \in \text{mod}(\delta)$. If K consists of the models of a set of evidence sentences—which is appropriate to the case of background knowledge that comes entirely from previous observations—then K will be compact in the evidence.[11] Let us say that ψ is *strongly not finitely verifiable* (respectively, *falsifiable*) relative to K just in case for any $w \in K$ and any finite set E of evidence sentences, if $w \in \text{mod}(E)$ and $w \in \text{mod}(\psi)$ (respectively, $w \in \text{mod}(\neg\psi)$), then there is a $w' \in K$ such that $w' \in \text{mod}(E)$ but $w' \notin \text{mod}(\psi)$ (respectively, $w' \notin \text{mod}(\neg\psi)$).[12] If there is a learning function F that reliably identifies the truth value of ψ relative to K, then ψ cannot be both strongly not finitely verifiable and strongly not finitely falsifiable relative to K. For suppose on the contrary that ψ is both. Then ψ is true in some members of K and false in others. Start with a $w \in K$ for which ψ is true and begin feeding F evidence sentences true in w. If F is reliable, it must at some point start conjecturing ψ. But since ψ is strongly not finitely verifiable, there is a $w' \in K$ in which all the evidence seen by F so far is true but in which ψ is false. If F is reliable, it must at some point start conjecturing $\neg\psi$ after being fed enough additional evidence true in w'. But since ψ is strongly not finitely falsifiable, there will be a $w'' \in K$ in which all the evidence seen by F so far is true and in which ψ is true as well. If F is reliable, it must at some point start conjecturing ψ after being fed enough additional evidence true in w''. Continuing in this way, we get an infinite sequence of evidence sentences for which F doesn't converge. Since K is compact in the evidence, there will be a member of K in which all the elements of the evidence sequence are true. Thus F isn't reliable after all.

Combining this result with the above discussion, we can conclude that no matter how complex the quantificational structure of ψ, the Bayesian is probabilistically certain that the actual world belongs to a narrow enough $K \subset \mathrm{Mod}_{\mathscr{L}}$ such that if K is compact in the evidence, then either there is a finite collection E of evidence sentences that are possibly true (there is a $w \in K$ such that $w \in \mathrm{mod}(E)$) and whose truth guarantees the truth of ψ (for any $w \in K$, if $w \in \mathrm{mod}(E)$, then $w \in \mathrm{mod}(\psi)$) or else there is a finite collection E of evidence sentences that are possibly true and whose truth guarantees the falsity of ψ. In the former case this means that to be a Bayesian I must either have probabilistic certainty to begin with that observation will never yield the conjunction $\&\, E$ of the sentences in E ($\mathrm{Pr}(\&\, E) = 0$) or else I lack such certainty and set $\mathrm{Pr}(\psi/\&\, E)) = 1$ so that upon observing that the conditions $\&\, E$ obtain, I (being a conditionalizer) raise my degree of belief in ψ to 1. If I were to do astrophysics, however, I would not want to be committed in advance to certainty that some finite condition $\&\, E$ about galaxies, stars, and planets would never be found to obtain, nor would I want to be committed to raising the probability of 'For every galaxy, there is a star with planets, and for every such planet, there is a moon such that ...' to 1 if I were to find that $\&\, E$ did obtain. My qualms about the latter case are exactly parallel.

When we concentrate on a $K \subseteq \mathrm{Mod}_{\mathscr{L}}$ that is measurable by Bayesian lights, we can bring into play a much more telling result due to Kelly

(1990). Recall from chapter 6 that the measurable subsets of $\text{Mod}_{\mathscr{L}}$ lie in the σ field generated by sets of the form $\text{mod}(\varphi)$, where φ is a sentence of $\mathscr{L}$. Thus, a measurable K will be of the form $\text{mod}(\Delta)$, where Δ is a finite or countably infinite set of sentences or, more generally, a finite or countable union of such sets of models. In section 5 we saw that there is a formal-learning function that reliably identifies the truth of ψ for all $\text{Mod}_{\mathscr{L}}$ just in case ψ is equivalent in $\text{Mod}_{\mathscr{L}}$ to a verifutable sentence. Kelly strengthens this result by showing that there is a formal-learning function that reliably identifies the truth of ψ for $K = \text{mod}(\Delta)$ just in case ψ is equivalent in K to a verifutable sentence.

Once again consider a ψ with a quantifier structure as complex as you like, and apply the Bayesian convergence-to-certainty result to conclude that there is a K of measure 1 such that the Bayesian learner converges to 1 (respectively, 0) in any $w \in K$ in which ψ is true (respectively, false). And again, since a Bayesian learner is matched by a formal learner, it follows from Kelly's result that for K of the form $\text{mod}(\Delta)$, the Bayesian converges to certainty because he is probabilistically certain *ab initio* that the world is such that the multiple quantification involved in ψ collapses so that ψ is equivalent to a verifutable sentence.[13] Popperians who are willing to enlarge their focus from refutation to verifutation will feel partially vindicated.

7   In What Sense Does a Formal Learner Learn?

Having used formal-learning theory to unmask some of the hidden assumptions of Bayesianism, one finds it natural to wonder whether formal-learning theory can be thought of as a replacement for Bayesianism. I think the answer is no. Formal-learning theory is best seen as directed toward a form of knowledge acquisition that, to use the frayed but still useful nomenclature, is closer to knowledge how than to knowledge that. Thus, for example, the seminal work of Osherson, Stob, and Weinstein reported in *Systems That Learn* (1986) was motivated by the case of language learning, where to learn a language means (roughly) to develop the ability to correctly identify the grammatical strings of the language. Nevertheless, formal-learning theory can also be put to work in the service of knowledge that, at least if we accept the tradition, going back as far as Plato, that holds that knowledge is reliably acquired true belief. For the reliability clause of the definition of a formal-learning function dovetails nicely with this tradition, since it guarantees that the agent following the rule would still have arrived at the (possibly different) truth even if the world had been different and she had been presented with different evidence.

I do not wish to review here the currently lively debate about reliabilist conceptions of knowledge.[14] So I will confine myself to two remarks. First, I am not generally predisposed against reliabilist conceptions. It seems to me, for example, that a reliabilist account of perceptual knowledge of the external world is probably the preferred account: I know that there are three books on the desk I am using to write this passage because this proposition is true, because I believe it is true, because my belief is the result of a belief-fixing process that is reliable across the normal range of perceptual situations, and because this case is in the normal range. Second, however, it seems to me just as evident that in the realm of scientific hypotheses and theories, reliabilist considerations are relevant to the sorts of knowledge claims of interest to scientists and philosophers of science only insofar as these considerations can be given a Bayesian interpretation. Indeed, it is not knowledge but justified belief that is the important issue. And for both the working scientist and the philosopher seeking to understand the methodology of science, the only kind of justification of interest is *articulable justification*: the evidence that forms the basis for the justification must be capturable in the form of a proposition, and the way in which the proposition supports the hypothesis in question must be capable of being made explicit. The burden of the argument of the preceding chapters is that while all may not be well with orthodox Bayesianism, it nevertheless offers the best available means for assessing the bearing of evidence on hypotheses.

Let me illustrate. Suppose (not so implausibly) that the original hand of nature endowed Einstein with a procedure that, given enough of the relevant data, eventually settles on the true theory of gravity. Then on the reliabilist conception, Einstein knew that GTR is true (assuming that it is) when his procedure settled on this theory and he came to believe the theory as a result. I will not quibble with the ascription of the term 'knowledge'. The real issue is whether Einstein and others were justified, either in November 1915 or at a later date, in according GTR a high degree of belief. Reliabilist considerations may, of course, be relevant to this issue, but to tell whether and how they are relevant requires viewing them through the lens of Bayesianism. Einstein's spectacular track record of successes is a relevant piece of information, and its relevance gets expressed, say, in terms of the prior probability assigned to GTR. If in addition Einstein or others can explicitly formulate a formal-learning rule that provably converges on the true theory of gravity and that, when fed the positive results of the three classical tests, conjectures GTR, that is additional relevant information that influences the posterior probability of GTR. But the question arises as to how well the three classical tests separately and jointly support GTR, and here there is no systematic means available to answer this and similar questions, save for Bayesianism.[15]

8 Putnam's Argument Revisited

This chapter began with a review of Putnam's 1963a argument and then moved rapidly through a series of considerations that have taken us a good distance from Putnam's original concerns. In closing the chapter, I want to return to Putnam's construction. Putnam had the avowed purpose of convincing Carnap that his approach to inductive logic ought to be abandoned. Putnam's indictment levels the charge that an approach based on c functions (or more generally, a minimally recursive Pr function) cannot capture the judgments of a "good" or "ideal" inductive judge. The charge has two parts, the first of which I have already rehearsed, namely, that once a scientist announces that she is using a particular minimally recursive Pr function, we can diagonalize to produce an effective hypothesis about the distribution of redness (say) among the sequence of individuals $a_1, a_2, \ldots,$ whose truth the scientist will never discern by peering through her probability window. The second part of the charge consists in the claim that if someone proposes this probabilistically unlearnable hypothesis and if its predictions are borne out without exception, then eventually the ideal judge will base her predictions on the hypothesis and will continue to do so unless the hypothesis fails.

There are two things to be said in response to Putnam's charge. The first begins with the remark that the charge has the most sting when posed in the form suggested in section 5; namely, there is a class of hypotheses whose truth the judge can reliably identify using a recursive formal-learning rule but not by using a minimally recursive Pr function. However, the sting of the charge is considerably dissipated by the discussion at the end of section 5. For if the conjectures made there are correct, the existence of a reliable formal learner for $\text{Mod}_{\mathscr{L}}$ is matched by the existence of a reliable

Bayesian learner, albeit one using a nonrecursive Pr function. Using such a function may not create insuperable problems in practical applications if the brain of the judge is a physical system that can analogue-compute Pr values.[16]

Second, while Putnam's charge has the merit of highlighting the potential of a tension between two aims of inductive methods, discovery and justification, it slights the second aim. Grant for the sake of argument that after experiencing nothing but positive instances, the ideal judge will eventually base her predictions on H_D and will continue to do so until H_D fails. We still want to know how long "eventually" is. Specifically, how many positive instances are needed to warrant acceptance in the sense of willingness to bet at specified odds that the next instance (or all future instances) will also be positive? Putnam's inductive judge is silent on this matter and apparently must remain silent until she learns how to speak in the Bayesian idiom.

It is useful to turn from these metareflections on methods to the point of view of a particular inductive agent. Given Putnam's avowed purpose of getting the Carnapians or, more broadly, the Bayesians to abandon their way of doing business, it seems fair to consider his complaint from the point of view of an inductive judge who is a practicing Bayesian using a minimally recursive Pr function. It helps to further distinguish two scenarios. If this ideal inductive judge is indeed ideal, we may suppose her to be logically omniscient (chapter 5). She will therefore have explicitly formulated Putnam's diabolical hypothesis $H_D((\forall i)(Pi \leftrightarrow \neg Ra_i))$ and will have realized that her prior probability assignments make it impossible for this hypothesis to achieve and maintain a high probability, even in the sense of instance confirmation. But being a good Bayesian and having the courage of her convictions, she sees no awkwardness in deviating from Putnam's prescription for an ideal inductive judge. On the other hand, we may suppose more realistically that our judge, like an actual scientist, is far from achieving logical omniscience and either does not explicitly formulate Putnam's hypothesis *ab initio*[17] or else does not realize that her probability assignments preclude her from learning that it is true. When a fellow scientist does explicitly propose the hypothesis and notes that it has been borne out without exception for thousands of cases, our judge may reassess her original probability assignments so as to make the hypothesis confirmable. As argued in chapters 5 and 8, this is a real and important phenomenon in science, but it is one that derives from the failure of logical omniscience and has nothing to do per se with Carnapian measure functions, effective computability, and diagonalization.

It might be complained that the appeal in the first part of this response to the agent's having the courage of her convictions is just so much braggadocio, for actual inductive agents, even would-be Bayesian agents, need not behave in this way. Just as the introduction of previously unformulated hypotheses can cause non-Bayesian shifts in the degree-of-belief function (see chapters 5 and 8), so the acquisition of new evidence can cause the reassessment of the original probability assignments. In the case in point, seeing the pattern of reds and nonreds conform unerringly to the probability-zero pattern of $(\forall i)(Pi \leftrightarrow \neg Ra_i)$ may eventually lead the agent to shift to a different degree of belief function, or so the objection would hold. To admit the force of this objection is to start down a road that leads very quickly to the evisceration of Bayesian confirmation theory.

All things considered, can we not say that Putnam's query does at least produce a pressure for a form of Bayesianism more circumspect than traditional Bayesian personalism? Specifically, if there are Pr functions that correctly converge to 1 or 0 on all those true hypotheses that are of interest and are learnable by means of the non-Bayesian formal-learning paradigm (suggested by Putnam's remarks) for every possible evidence sequence drawn from any $w \in \text{Mod}_{\mathscr{L}}$, should not the wise Bayesian operate with one of these Pr functions? All things being equal, the answer seems to be yes. But as usual, all other things may not be equal. Choosing a certain prior distribution may preclude universal learning, but this choice may have the compensating virtue of getting one very quickly to the truth for hypotheses that are taken to matter. Similarly, using a recursive Pr function may preclude the learning of some hypotheses, but these hypotheses may be regarded as relatively unimportant, and in any case, computability may make life so much more pleasant that the diminution in the range of learning may be a price worth paying. Such considerations have not received much attention in standard discussions of Bayesian methodology. It is a virtue of Putnam's query that it has forced us to pay attention to them.

9 Conclusion

The original exchange between Putnam and Carnap was inconclusive, and reading that exchange when it first appeared left one uncertain as to who had scored any significant point.[18] In the light of formal learning theory, however, Putnam's diagonalization argument can be construed so as to score at least one point against the Carnapian approach and, more generally, the Bayesian approach to induction. That score comes from examples of learning problems that can be solved by an effective formal learner but not by any effective Bayesian learner.

More important, in a sort of judo move in which the strength of Bayesianism is used against itself, formal learning theory reveals that the convergence-to-certainty theorem that links Bayesianism to truth and reliability is underwritten by dogmatism of a kind in the form of substantive a priori knowledge. We thus seem to be faced with a dilemma. On the one hand, Bayesian considerations seem indispensable in formulating and evaluating scientific inferences. But on the other hand, the use of the full Bayesian apparatus seems to commit the user to a form of dogmatism.

10 A Dialogue

Sol, a young philosopher of science from an Ivy League university, joins Sue and Sam, two colleagues from the University of Nevada at Las Vegas. Here we catch them at lunch at the Golden Nugget during a break from their gambling spree.

Sol   The games of chance in your wonderful casinos suggest to the studious mind a large field for investigation, especially in the area of probability and inductive inference.

Sam   You're right. Being studious by nature, I frequently read not only books about how to improve my odds at the tables but also more theoretical works concerned with the foundations of probability and ampliative inference.

Sol   Then perhaps you've looked at *Bayes or Bust?*

Sam   I have. Sue and I have spent many hours poring over this difficult and disturbing work.

Sol   I can readily understand why you say it is difficult, but why do you say it is disturbing?

Sam   The author convinces me that Bayesianism holds the best hope for constructing a unified and comprehensive account of scientific inference. But at the same time he also convinces me that Bayesian confirmation theory faces some thorny if not crippling difficulties. Isn't this your reading too, Sue?

Sue   Yeah.

Sol   Tell me more.

Sue   One worry concerns the status of Bayesianism. You can't view it as descriptive of the actual beliefs scientists hold. And in view of the recent results of formal-learning theory, you also can't view it as a normative theory. Those results imply that using the Bayesian apparatus commits the scientist to what look suspiciously like substantive assumptions about the world. For instance, when the Bayesian's background knowledge can be expressed in terms of a set of sentences, it follows that she must be probabilistically certain that a hypothesis with a quantifier structure as complicated as you like is equivalent to a "verifutable" statement, a statement that is a truth-functional compound of statements each of which can either be verified or falsified by finite data. For astronomical hypotheses such as 'For every galaxy, there is a star with planets, and for every such planet, etc.', such certainty seems to amount to substantive astronomical knowledge. While an astronomer might bring such knowledge to an investigation of some problem, it seems misguided to *require* that she start the inquiry with such knowledge.

Sol Well spoken, Sue. But surely there must be a way out for the Bayesians, who always seem to have multiple escape routes from any objection.

Sam If I remember correctly, the result Sue refers to depends on the assumption that the agent's degree-of-belief function satisfies countable additivity. Some Bayesians are already leery of this requirement for other reasons, and they may see this difficulty as a further reason for rejecting this requirement.

Sol What say you to that, Sue?

Sue The difficulty I see with this way out is that Bayesians will no longer be able to prove their famous convergence-to-certainty and merger-of-opinion theorems. But even if this problem is solved, there remain further challenges to the normative status of Bayesianism. For instance, the rule of belief change by conditionalization cannot be seen as a rationality constraint; or to put it more cautiously, this rule does not have the same firm grounding in Dutch-book considerations as do the axioms of probability. Even more fundamentally, the norms of orthodox Bayesianism require logical omniscience, a demand that actual scientists with their computational limits cannot hope to satisfy. Thus, insofar as 'ought' implies 'can', the 'oughts' of Bayesianism are suspect.

Sol I seem to recall from *Bayes or Bust?* that there are two relevant senses of logical omniscience involved here. Can you remind me of what they are?

Sue Yes. The first sense requires that the Bayesian agent recognize all logicomathematical truths, in particular, truths about logicomathematical implications. You can easily appreciate why actual scientists cannot live up to this requirement. Not even Einstein himself recognized all of the implications of, say, his own theories of relativity, and what knowledge of this sort he did have was often gained only through difficult calculations.

Sam Before we continue, I think we need to clarify the sense in which Bayesianism requires logical omniscience. It is part of the probability calculus that if $\models A$, then $\Pr(A) = 1$, and in that sense, degrees of belief that conform to the probability calculus must respect logical truth. But normally, statements of the form '$\models A$' are not themselves members of the set of propositions to which probabilities are assigned, so it is not required that if $\models A$, then $\Pr(\models A) = 1$.

Sol I am not sure where your observation is leading us, Sam.

Sam To elaborate Sue's example, let A be (GTR $\rightarrow P$). [Sam writes this down on a napkin.] 'GTR' stands for Einstein's general theory of relativity, and P is the statement that the anomalous advance of Mercury's perihelion is 43 seconds of arc per century. At first Einstein did not recognize that this conditional is a theorem of the appropriate mathematical system for GTR. Nevertheless, we may plausibly take him to have assigned a probability of 1 to this conditional. For he knew P to be true, and so $\Pr(P)$ equaled 1 for him. Thus, if his degrees of belief respected some transparent truths of propositional logic, $\Pr(\text{GTR} \rightarrow P)$ must also have been 1 for him.

Sue Very clever, Sam. But there will be other cases where your observation does not apply. For example, if P is a prediction of the gravitational-lens effect, then presumably in 1915 Einstein set $\Pr(P) \neq 1 \neq \Pr(\text{GTR} \rightarrow P)$, even though GTR does in fact entail this effect. Thus in 1915 Einstein's degrees of belief violated the probability calculus. Moreover, if we concentrate on the examples of your sort where $\Pr(P) = 1$, then the problem of old evidence rears its ugly head.

Sol This conversation is going too fast for me. I do want to hear about the problem of old evidence, but let me suggest that we first concentrate on the problem that Sue initially raised; namely, even the best of scientists, Einstein included, cannot be expected to have "coherent" degrees of belief, because they are not logically omniscient in the sense that Sue gave. And I would add that scientists do logical as well as mathematical learning, so that it would seem reasonable to demand that a realistic Bayesianism should account for both forms of learning.

Sam Daniel Garber has shown how logical learning can be accommodated by the model of learning as conditionalization if we work in a language built up like this. The basic sentences consist of atomic sentences.

Among them are 'GTR' and 'P' from our working example, plus sentences of the form '$A \vdash B$', where A and B are atomic. [Sam is again writing on his napkin.] Then we take arbitrary finite truth-functional combinations of the basic sentences. Within this language, '$\vdash$' is a primitive, though extrasystematically it will be interpreted as the form of logicomathematical implication appropriate to general relativity. With '$\models$' interpreted appropriately to the truth-functional language I just sketched, it will not be true either that $\models (\text{GTR} \rightarrow P)$ or that $\models (\text{GTR} \vdash P)$. Hence, for a probability function defined on this language, there is no incoherence in setting $\text{Pr}(\text{GTR} \rightarrow P) \neq 1 \neq \text{Pr}(\text{GTR} \vdash P)$. And when the Bayesian agent does learn that GTR entails P, he can represent this new knowledge in terms of a shift from Pr to $\text{Pr}'(\cdot) = \text{Pr}(\cdot / \text{GTR} \vdash P)$.

Sue You have a penchant for the clever, Sam. But it is not clear to me how these clever formal results resolve our worries, since the results rest on the pretenses, which the Bayesian agent well knows to be false, that the sentence 'GTR' is atomic and that a possible world is just an assignment of truth values to the basic sentences of the toy truth-functional language. And there remains the nagging doubt that this apparatus contains lurking inconsistencies, or at least lurking inadequacies, such as the inability to handle conditional proofs.[1]

Sol All this is very interesting, but I fear that it is leading us off on a tangent. To continue us down the main thread of my question, let me ask you to remind me of the second sense of logical omniscience and how its failure can be accommodated in a realistic Bayesianism.

Sam Logical omniscience in the second sense requires that the agent parse all of the possibilities at the start. And the standard way to take note of the failure is to introduce a catchall hypothesis H_C. This hypothesis says, in effect, that some hypothesis is true that we cannot now explicitly articulate and that lies beyond the hypotheses articulated to date.

Sue It is one thing to pay lip service to the failure of this second form of logical omniscience by introducing the catchall, and it is quite another to show that the Bayesian apparatus can continue to function after this introduction. For example, the denominator on the right-hand side of Bayes's theorem is typically written as

$$\text{Pr}(E/K) = \sum_i \text{Pr}(E/H_i \,\&\, K) \times \text{Pr}(H_i/K).$$

[Here Sue takes to writing on a napkin.] If the summation is taken to include the catchall, as apparently it must be, then the Bayesian agent needs to evaluate the factor $\Pr(E/H_C \& K)$. But such an evaluation would of necessity seem to be ill informed, since H_C stands for unexplored territory.

Sam   Can't we avoid this embarrassment by concentrating on the ratios $\Pr(H_i/E \& K)/\Pr(H_j/E \& K)$, where H_i and H_j are two of the explicitly formulated hypotheses? For then the troublesome factor $\Pr(E/K)$ is canceled out.

Sue   This is a move recommended by Wesley Salmon. It works if what we are trying to do is to choose among the explicitly formulated hypotheses. But this move seems most un-Bayesian, since Bayesian epistemologists don't want to choose hypotheses, in the sense of accepting or rejecting them, but rather aim to probabilify them. Moreover, the Bayesian needs the probability values of the hypotheses, not just the ratios of these values, if she is to supply one of the essential ingredients needed in decision making under uncertainty. Nor can the factor $\Pr(E/K)$ be ignored if the Bayesian hopes to explain judgments to the effect that one piece of evidence confirms a hypothesis better than another.

Sol   Another reaction would be to get a better fix on $\Pr(E/H_C \& E)$ by actively exploring the region of possibility space for which H_C is a name.

Sue   This option is discussed in *Bayes or Bust?* in terms of some very interesting case studies. But the results of such an exploration cannot be modeled by any of the standard Bayesian paradigms.

Sam   Why doesn't the catchall serve us here? As each new explicitly formulated hypothesis is introduced, some of the probability previously attached to H_C is transferred to the new hypothesis. This is hardly a radical departure from standard Bayesianism.

Sue   As a matter of actual historical fact, the introduction of new hypotheses can result in a much more radical redistribution of probabilities than your simple model of shaving off probability from the catchall would suggest. For example, Einstein's introduction of his special theory of relativity was accompanied by a redistribution of probability over the previously formulated theories. In particular, the introduction resulted in a subtraction of probability from classical electron theories.

Sam But this is an example of a major scientific revolution. Why shouldn't the Bayesian candidly admit that his apparatus fails to apply to the discontinuities encountered as one crosses the frontiers of a scientific revolution while maintaining that his doctrine applies in normal science?

Sue If a redistribution of probabilities not equivalent to shaving off from the catchall is taken to be a definition of a scientific revolution, then such revolutions occur with monotonous frequency, and the applicability of Bayesianism threatens to shrink to the vanishing point.

Sol If I may change the subject somewhat, isn't the problem of old evidence linked to the issues we are now discussing?

Sue You are quite correct, Sol. To revert once more to our earlier example, Einstein knew the value of the anomalous advance of Mercury's perihelion (P) long before he invented his general theory, and in this sense it was old evidence. When he formulated the theory in November 1915, he had to shift to a new probability function $Pr_{new}(\cdot)$ either by shaving off from the catchall or by some more radical means. However this was accomplished, $Pr_{new}(GTR/P) = Pr_{new}(GTR)$ if $Pr_{new}(P)$ remained equal to 1, as it presumably did for Einstein and others. Thus on the Bayesian analysis it would seem that P does not incrementally confirm GTR, which flies in the face of strongly held intuition.

Sol If I understand you correctly, Sue, the failure of logical omniscience in the second sense is responsible for the sticky version of the problem of old evidence. For example, the fact that the data P for perihelion advance is now old news does not by itself prevent us from saying that P confirms GTR. For imagine that GTR had been explicitly formulated at the beginning of inquiry and that no nonlogical learning has taken place in the meantime. Then Pr_{new} arises from a series of conditionalizations on empirical evidence statements that have been learned along the way, and we can simply trace back to the stage when P was first learned to find an incremental confirmation of GTR by P.

Sue   Correct. But when logical omniscience in the second sense fails, it seems that to accommodate the assertions that the perihelion phenomenon confirms GTR, that it confirms GTR more than either the bending of light or the red shift, etc., the Bayesian must appeal to counterfactual beliefs, for example, to degrees of belief that Einstein and others would have had if they hadn't known the evidence *P* prior to the formulation of GTR. But it is far from evident that such appeals will validate our clear and strong intuitions on this matter.

Sam   Earlier I sketched Garber's device for handling the failure of the first form of logical omniscience. I suggest that the same device can be used here to resolve the problem of old evidence. When Einstein proved in November of 1915 that his GTR entailed the observed advance of the perihelion of Mercury (*P*), he did a piece of logicomathematical learning. In Garber's system it is possible to have that $\Pr(\text{GTR} \vdash P) < 1$ and thus that $\Pr(\text{GTR}/\text{GTR} \vdash P) > \Pr(\text{GTR})$. So learning the entailment can boost confidence in the theory.

Sue   This may be the solution to some problem, but it is not a solution to the original problem of old evidence. That problem is to account in Bayesian terms for the generally accepted notion that the previously known evidence *P* has strong confirmatory value for GTR.

Sol   As interesting as this cluster of problems is, I think it is time to move on. Let me ask Sue, who seems to be the most pessimistic among us, what it is that worries her the most about the Bayesian approach to confirmation.

Sue   My major worry transcends the problems we have been discussing so far and concerns the significance of the little numbers the Bayesians assign to hypotheses. I of course endorse Bayesian personalism as part of a theory of individual decision making under uncertainty. When a person decides which course of action to take, it is *her* assessments of the utilities and the probabilities of the outcomes that count. But when it comes to assessing how well a scientific hypothesis is confirmed, it is harder to see how these little numbers, which measure how warmly the person feels about hypotheses, can have the kind of rational and objective significance supposed to be part of scientific inference.

Sam   You have raised a major problem that we cannot hope to resolve today. But to make a start, I urge that we divide the problem. As the author of *Bayes or Bust?* notes, one threat to objectivity arises from the conjunction of the facts that confirmation values seem to depend on the choice of language or possibility set and that this choice is in significant part a pragmatic affair. He seems to feel that the relativism involved here is inevitable in any nonfoundational approach to knowledge and also that it is relatively harmless. I am not at all sure that I agree with him. But here I want to concentrate on the second aspect of the objectivity problem that arises once we have settled on a language or a possibility set. In this case, I suggest, part of the answer to your worry is provided by the convergence-to-certainty and merger-of-opinion theorems, which the Bayesians are justly proud of. The latter shows that the Bayesian method leads to objectivity in the sense of intersubjective agreement, and the former shows that the method reliably leads to truth.

Sue   Don't forget what we learned near the beginning of our conversation: the very apparatus that generates these results commits the Bayesian to making substantive knowledge claims at the very beginning of inquiry, a commitment that looks very much like a form of dogmatism. The Bayesians therefore seem to be impaled on a dilemma: dogmatism or else no reliable connection to the truth. Furthermore, the theorems in question do not apply to theoretical hypotheses that are underdetermined by the data. And the theorems, where they do apply, are in the form of limit results, which in general are unaccompanied by any useful estimates of how fast the convergence takes place.

Sam   Could we not argue on evolutionary grounds that for various propositions our degrees of belief rapidly converge to reliable estimates of objective probabilities or relative frequencies, or else we would not have survived? In this way the little numbers take on an objective significance even in the short and medium runs.

Sol   In some instances this may be plausible. But it is hard to see how to regard the degrees of belief we assign to the esoteric hypotheses of modern science as estimates of frequencies, and it is hard to understand why accuracy in such estimates confers evolutionary advantages. And as noted in *Bayes or Bust?* even for lowly observational hypotheses that have direct survival value, we may survive and prosper not because of the accuracy of our estimates of relative frequencies but because we have evolved the ability to tolerate or to creatively cope with the consequences of our inaccurate estimates.

Sue I agree with you, Sol. But I don't think that I have made sufficiently clear the core of my worry about the significance of the little numbers the Bayesians use. Grant for the sake of argument that the Bayesians can represent features of confirmation by means of their little numbers. What still nags at me is the sense that underneath the little numbers are factors that need to be analyzed in non-Bayesian terms. Thus, for example, I grant that the confirmation a hypothesis receives from a positive instance can be represented in terms of an incremental increase in probability. But the underlying reason that there is confirmation is to be found in the non-probabilistic structural relation between the evidence statement and the hypothesis. Hempel, Glymour, and others have tried to analyze this relation, and although they may not have succeeded, my feeling is that they are onto a crucial feature of confirmation that does need spelling out. As another example, we all know that in science and everyday life there are many possibilities that we don't take seriously. Bayesians can represent this by saying that low prior probabilities are being assigned. But I think that this is representation and not explanation. For in some cases we can back up our dismissal of various possibilities by giving plausibility arguments, and on pain of circularity or regress, these arguments cannot be given a Bayesian analysis. I could go on and on in this manner.

Sam I know you could, Sue, because I have heard you do it on many occasions. And for Sol's benefit, I will repeat what I always say in response. If you are correct and Bayesianism is just a tally device used to keep track of a more fundamental process, then you ought to be able to produce a non-Bayesian analysis of confirmation and induction. But I hardly need to remind you that what you have produced consists mainly of promissory notes.

Sol And I would add that it doesn't seem entirely fair to say that Bayesianism gives representation without explanation. For instance, it seems to me that Bayesianism does explain why (the problem of old evidence aside) hypotheticodeductive evidence is confirmatory, and why (relative to plausible assumptions about the sizes of reference classes) the observation of a black raven confirms 'All ravens are black' more than does the observation of a white shoe.

Sue   True enough. Bayesianism is not without its successes. But it doesn't seem to live up to its billing as the be-all and end-all of scientific inference. As we Las Vegans would say, Bayesian confirmation theory pretends to be a main attraction when it is really just a lounge act.

Sam   I think that is being a bit unfair, Sue. In view of the lack of competition, I think it is more accurate to say that Bayesian confirmation theory is a main attraction but admittedly one that has not quite gotten its act worked out.

Sol   My brain is reeling from our discussion. My mind, like a cloud momentarily illuminated by a lightning flash, is sparking all sorts of strange, crude ideas. Let us agree to meet again tomorrow at the same time to discuss these matters further.

Sue   Agreed. And when we meet, let us confront more directly what I take to be the overarching problem that has emerged from our discussion. I can set the problem up by noting that there is a strong consensus in the scientific community to the effect that on the basis of the available evidence the earth is much older than the ten to twenty thousand years the creationists posit, that Velikovsky's *Worlds in Collision* scenario is bunk, that there is no ether, and that space and time are relativistic rather than absolute, etc. Now one attitude we can take is that an account of scientific inference that does not underwrite these firm judgments is at best incomplete and at worst bankrupt. The contrasting attitude of some of the Bayesian high priests is that these beliefs may not be so rational and objective as we would like to think. For surely the available evidence does not force a tight merger of opinion on these matters for a maximal class of equally dogmatic prior-belief functions, and it seems a reasonable surmise that the merger will materialize only if the starting class is very circumscribed. If we hew to the former attitude, we will need to supplement the principles of Bayesianism or else to give a different account of the rationality and objectivity of our shared beliefs. The Bayesian priests would respond that their doctrine supplies the principles of inductive logic, that to go beyond this logic one must invoke substantive assumptions, and that such assumptions cannot be part of a theory of rationality. Our discussion has revealed that the priests are wrong on this point. For a Bayesianism worth its salt must invoke a rule of conditionalization and must use countable additivity to prove the reliability of the method in the form of convergence to the truth, but as we have seen, these invocations involve substantive assumptions. The way is thus open for developing a more appealing account of scientific inference by adding other substantive assumptions. But what should they be? And what shape should the account take? I suggest that we start with these questions when we reconvene.

Sam I cannot agree to your slanted version of the agenda. We should keep a more open mind, for example, by keeping ourselves open to the possibility that the explanation of consensus in a scientific community typically has a large historical and sociological component.

Sue By all means keep an open mind, but not so open that your brain falls out.

Sam In referring to a sociological component, I am not suggesting anything as crude as brainwashing. Rather, I am pointing to the commonplaces that a scientific community is not a community unless there is a consensus about the shape of the explanatory domain, about what instruments are reliable, etc., and that the training of neophyte scientists consists in part in transmitting to them this shared wisdom. At least some of this wisdom can be expressed in terms of degrees of belief, and so, becoming part of the community means aligning one's degrees of belief with the consensual degrees of belief of the community.

Sue Sure. But we come back again to the question of from whence arise these consensual degrees of belief. If they have arisen from a wide class of initial belief functions via conditionalization on the accumulated evidence of observation and experiment, then the indoctrination of neophyte scientists is just a shortcut method of learning. But we have agreed that this is not likely to be the explanation, and to the extent that it is not the explanation, the indoctrination of neophyte scientists begins to look like brainwashing.

Sam Throwing terms like 'brainwashing' around doesn't advance the discussion. I am suggesting, for example, that the explanation of why the present physics community assigns a high degree of belief to the proposition that space-time is relativistic rather than Newtonian has a large historical and sociological component. Once the details of this component are laid out—and the laying out would involve telling much of the history of physics from Galileo to Einstein—you will see that there is nothing sinister in what I am proposing.

Sue Sinister or not, it seems muddleheaded to me. I thought that you were favorably inclined toward Bayesianism, but your last remarks make me wonder about where your loyalties really lie.

Sol I fear that our conversation has degenerated into bickering. Let's discuss the agenda after we have given it some thought. In the meantime, I want to get back to the casino.

Sue Try the slot machines nearest the door. They have higher odds of paying off.

Notes

Introduction

1. See Barnard's (1958) biographical sketch of Bayes.

2. See Stigler 1982, 1986 for an account of the influence of Bayes on the development of statistical thinking.

3. For recent commentaries on Bayes's essay, see Dale 1982; Edwards 1978; Gillies 1987; Hacking 1965, 1970, 1975; Pearson 1978; Pitman 1965; and Stigler 1982, 1986.

Chapter 1

1. Answer: 25, as first shown by De Moivre.

2. Page references are to the *Biometrika* edition of Bayes's essay.

3. Actually, Bayes simply says 'table', not 'billiard table.' But the tendency to use the latter has been so strong that few commentators have been able to resist it. Nor can I.

4. Rule 1 was composed by Bayes. Rules 2 and 3 and the appendix were supplied by Price. We do not know to what extent Price relied on Bayes's notes and to what extent his additions are original.

5. Of course, Bayes did not use integral notation but expressed his results in terms of areas under curves and ratios of areas. Modern notation will be used here.

6. Because belief takes propositions as its object, I will assign probabilities qua degrees of belief to propositions. The point about Bayes's definition of probability now becomes that the formula $\Pr(A) = x$ makes sense only if what proposition A asserts can be ascertained to hold or to fail. For ease of presentation, I will often speak of A as an event rather than the proposition asserting that the event occurs.

7. One could take the point of view that being bilked by a bookie who catches the violator of the probability axiom in a no-win, must-lose situation is only window dressing. Rather, what the Dutch-book arguments really reveal is a structural incoherency of preferences in the form of a finite series of bets, each of which is preferred to the status quo while the package of bets is less preferred than the status quo. Viewed in this way, Bayes's proof of proposition 1 is a species of Dutch-book argumentation. See chapter 2 for a fuller discussion of Dutch book.

8. Shafer asks, "Why does Bayes, in the statement of his third proposition, refer to the two events he is considering as 'subsequent'?" He answers, "Imagine a situation in which it is not known beforehand which of two events A and B will happen (or fail) first. In such a situation we cannot say beforehand what the probability of B will be immediately after A happens; that probability will be one if B has already happened by then, zero if B has already failed by then, and something in between if B has not happened or failed. Saying that B is subsequent to A may be an attempt on Bayes's part to resolve this ambiguity" (1982, p. 1078). I disagree, since, given a degree-of-belief conception of probability, the probability of B need not be 1 if B has already happened and 0 if B has already failed. Bayes's reference to subsequent events anticipates his billiard-table model, discussed below.

9. Unless it is assumed that a B event does eventually occur, the E_j will not be events in the sense of contingencies that can be ascertained at some finite time to hold or fail.

10. See Geisser 1980 for an overview of the predictivist point of view.

11. The timing was right. The best available evidence indicates that Bayes reached his result sometime between 1746 and 1749. Hume's *Treatise* appeared in 1739, and his *Enquiry* in 1748. For more on this matter, see Dale 1986, Gillies 1987, and Zabell 1989.

12. For an illuminating discussion of the relation between Hume's problem of induction and symmetry principles, see Zabell 1988.

13. Klibansky and Mosser (1954, p. 234). The cordiality of the correspondence was undoubtedly one factor in Price's decision to tone down his criticism of Hume in the second and later editions of *Four Dissertations*. In the second edition Hume is lauded as "a writer whose genius and abilities are so distinguished, as to be above *my* commendation" (1768, p. 382).

14. For an analysis of Price's criticism of Hume's view on miracles, see Dawid and Gillies 1989.

15. See also Edwards 1978. Another concrete illustration of the point is found in Fisher's (1956, pp. 123–126) "observations of two kinds." I am indebted to Sandy Zabell for this reference.

16. Dale seems to have been overly influenced by Karl Pearson's professed puzzlement over this case; see Pearson 1978, pp. 365–369.

17. From *Biometrika* 45 (1958): 296–315.

Chapter 2

1. Levi (1980) draws a subtle distinction between such temporal or dynamic conditionalization and what he calls confirmational conditionalization. The latter is atemporal in that it is a constraint on the agent's confirmational commitments at time t. It requires that the agent relate via conditionalization his current belief states to the hypothetical belief states that can arise by accepting new evidence. Levi himself accepts confirmational conditionalization but rejects temporal conditionalization, whereas Kyburg (1974) rejects both. The views of these authors are recommended to the reader for consideration, but they will not be discussed here.

2. This example is taken from Richard Jeffrey and Brian Skyrms.

3. For other approaches to conditionalization, see Field (1978) and Garber (1980).

4. Thus, if one wishes to be pedantic, a Bayesian probability space is a triple $(\mathcal{W}, \mathcal{A}, \text{Pr})$, where $\mathcal{W}$ is a set of possible worlds, $\mathcal{A}$ is a set of sentences or propositions, and Pr is a map from $\mathcal{A}$ to $\mathbb{R}$ satisfying the probability axioms.

5. In particular, if A is a tautology, then $\models A$.

6. See, however, Seidenfeld and Schervish 1983 for the problems this causes for Savage and de Finetti.

7. If the maximum odds an agent is willing to take *on* a proposition are less than the minimum odds she is willing to take *against* the same proposition, then Dutch book cannot be made against her. The resulting calculus of belief involves either subadditivity or interval valued degrees of belief. See Smith 1961 and Williams 1976.

8. Howson (1990a) attempts to overcome some of the above mentioned difficulties by focusing on some idealized contexts where there is a natural connection between subjective probabilities and propensities to bet and where the Dutch-book construction is sound. Howson seems to feel that although the conditions needed to run the Dutch-book construction fail outside of these idealized contexts, still "the constraints imposed as a consequence of Dutch Book considerations generalize quite naturally out of these simple contexts, for no other reason than because probability is a general guide, invoked impartially in *all* contexts" (p. 8). I leave it to the reader to evaluate the plausibility of this claim. Tim Maudlin (private communication) has noted that a thoroughgoing Bayesian ought to take into account the probability that the bet will be paid off. But when this is done, the connection between degree of belief and maximum betting odds may be severed. For example, I may assign a low

probability to the proposition that the world will end tomorrow, but since if it does end, I won't have to pay the bet, I am willing to bet against the proposition at any odds.

9. Suppose that an agent's utility U is a function from wealth to $\mathbb{R}$ and that U is twice differentiable. Economists typically assume that rational economic agents display both *non-satiation* and *risk aversion* for all levels of wealth (see Arrow 1971). In terms of U, these assumptions amount respectively to $U'(w) > 0$ and $U''(w) < 0$ for all w. To understand the economic implications of risk aversion, let X be a random variable that represents a risky asset in the minimal sense that X has a nondegenerate probability distribution, and let $E(X) = r^*$. From $U'' < 0$ and Jensen's inequality it follows that $E(U(X)) < U(E(X)) = U(r^*)$. Furthermore, if we define the cash equivalent of the risky asset X to be the amount of cold cash $\$r^{**}$ such that $U(r^{**}) = E(U(X))$, then $r^{**} < r^*$. For again by Jensen's inequality, $U(r^{**}) < U(r^*)$, and the result follows from nonsatiation.

10. Here 'estimate' must be taken in some primitive sense rather than in the probabilistic sense of expected value (see Shimony 1988).

The problem of the appropriate choice of reference class has bedeviled the frequency interpretation from the beginning. In this instance, however, the problem can be finessed (see van Fraassen 1983a).

11. Change of probability via conditionalization $\mathrm{Pr}_{new}(\cdot) = \mathrm{Pr}_{old}(\cdot /E)$ is not reasonable, and the Dutch-book arguments for the change reduce to nonsense unless E is the strongest proposition learned (see Mellor 1971).

12. The reader should be aware that Diaconis and Zabell (1982) assume, in effect, that the sensory stimulation relevant to the partition in question determines the probabilities over the elements of the partition independently of all prior experiences. Formally, this means that for any $\{F_i\}$, $\mathrm{Pr}_{FE}(E_i) = \mathrm{Pr}_E(E_i)$. Field (1978) wants to allow that the probabilities attached to the elements of a partition depend not only on sensory stimulation but also on prior probabilities.

13. Burks (1977) tries to seize the first horn.

Chapter 3

1. For a thoroughgoing criticism of HD methodology, see Glymour 1980. For a defense, see Horwich 1983, but see also Gemes 1990a.

2. Throughout this chapter it is assumed that standard first-order logic is operating.

3. Showing this is left as an exercise to the reader.

4. See Hempel's (1945) discussion of the Nicod criterion of instance confirmation; see also the discussion of the ravens paradox in section 3.

5. Thus, although $Ra \& Ba$ does not directly Hempel-confirm $Rb \to Bb$, it does Hempel-confirm it, since $Ra \& Ba$ directly confirms $(\forall x)(Rx \to Bx)$, which entails $Rb \to Bb$.

6. John Norton has provided a neat technique for constructing counterexamples. Suppose that an airplane has crashed in the jungle. Consider hypotheses that postulate that the crash took place in specified areas, and consider evidence statements that delimit the possibilities for the crash site. By representing these hypotheses and evidence statements on a Venn diagram, the reader can easily produce counterexamples to the special consequence condition and to each of the following seemingly plausible principles: (1) if E confirms H_1 and E confirms H_2, then E confirms $H_1 \& H_2$; (2) if E_1 and E_2 each confirm H and $E_1 \& E_2$ is self-consistent, then $E_1 \& E_2$ confirms H; (3) if E refutes $H_1 \& H_2$, then E does not confirm each of H_1 and H_2. The reader should also note that the Popper and Miller argument studied below in chapter 4 inadvertently provides another type of counterexample to Hempel's

special consequence condition, namely, $H \models H \vee \neg E$, but under minimal conditions, E incrementally disconfirms $H \vee \neg E$, even though it may confirm H.

7. Such uninteresting cases include those in which H can be written as $H_1 \vee H_2$, where H_1 is not a logical truth and its nonlogical vocabulary is purely observational.

8. In early versions of bootstrapping, the very hypothesis at issue was allowed to play the role of one of the auxiliaries in the bootstrap calculations. Edidin (1983) and van Fraassen (1983b) have argued that such macho bootstrapping is both undesirable and unneeded.

9. In this section I ignore Goodman's problems. Thus, assume that all the predicates are non-Goodmanized and "projectable."

10. An exercise for the reader: prove these facts.

11. Suppose that the background knowledge K^* specifies that the individuals a and c are random selections respectively from the class of ravens and the class of black things. It is subsequently found that a is black and that c is a raven. Object a was a potential falsifier of the ravens hypothesis H, while c was not. Does Ra & Ba give better confirmational value than Rc & Bc? The reader is invited to explore this question in the Bayesian manner. Interpret 'confirmational value' in the incremental sense, and then begin (as always!) by writing out Bayes's theorem to determine the conditions under which

$Pr(H/Ra$ & Ba & $K^*) > Pr(H/Rc$ & Bc & $K^*)$.

12. The ravens paradox remains one of the most contentious topics in all of confirmation theory, and it would be naive to think that my remarks will dissolve the controversy. However, I do hope that they serve to illustrate the fruitfulness of the Bayesian approach. For a sampling of some recent opinions on the ravens paradox, see Lawson 1985, Watkins 1987, French 1988, and Aronson 1989.

13. Similar examples can be worked out for van Fraassen's (1983b) semantic version of bootstrapping. To illustrate, take a theory T to be the closure under arithmetic operations of a set of linear equations, and take a data set E to be an assignment of values to the directly measurable quantities. Van Fraassen's conditions for bootstrap testing are as follows. E tests H relative to T just in case there is a $T_0 \subset T$ and an alternative E' to E such that (1) $T \cup E$ has a solution, (2) $T_0 \cup E'$ has a solution, (3) all solutions of $T_0 \cup E$ are solutions of H, and (4) no solution of $T_0 \cup E'$ is a solution of H. Take the axioms of T to be $H_1: A = B + C$, and $H_2: D = X$, where X is the theoretical quantity and A, B, C, D are the directly measurable quantities. Intuitively, H_2 should not be testable relative to T. But the formal definition is satisfied by taking T_0 to be $A - B - C + D = X$, E to be $\{A = 2, B = 1, C = 1, D = 3\}$, and E' to be $\{A = 2, B = 1, C = 2, D = 3\}$. It is no good to complain that E' contradicts H_1, since in bootstrap testing in general E' will contradict some consequence of T.

14. This idea was developed by Grover Maxwell and myself in the mid 1970s and circulated as a memo to members of the Minnesota Center for the Philosophy of Science. Similar ideas were developed by Franklin and Howson (1984) and Franklin (1990).

15. Elliott Sober (private communication) has pointed out that the idea that $Pr(E_n/\&_{i \leqslant n-1} E_i$ & $K)$ increases more slowly when the E_i are various is generally not correct outside of the context of HD testing. By making the likelihood factor equal to 1, the HD condition guarantees that the likelihood is independent of how varied the E_i are. But as Sober notes, in general this independence may fail, and when it fails, no conclusion can be drawn about the connection between the variety of the E_i and the value of ratios of the likelihood and prior likelihood factors without knowing more about the details of the case.

16. A more sophisticated Bayesian analysis of variety of evidence might exploit the notion of partial exchangeability (Diaconis and Freedman 1980). Roughly, there would be exchangeability (see chapter 4) within but not across different kinds. Variety of evidence would then involve instances from the different kinds. The confirmational virtues of such evidence is currently under study by Elizabeth Lloyd.

17. The connection between variety of evidence and eliminative induction is not a new idea; see Horwich 1982, chapter 6. The two aspects of variety of evidence mentioned above may be considered to belong to two different perspectives on evidence: posttrial evaluation (e.g., given the outcomes, how is information about their variety relevant to how much they boost the probability of the hypothesis?) versus experimental design (e.g., how is variety relevant to the design of experiments whose outcomes are most likely to boost the probability of the hypothesis?). A more detailed discussion of variety should pay careful attention to these perspectives. Here again I am indebted to Elliott Sober.

18. The symbolism $c(H, E)$ was Carnap's notation for the degree of confirmation of H on E; see chapter 4 for more details about Carnap's systems of inductive logic. Putnam thought that Carnap's inductive logic was subject to the following bind. "With respect to the actual universe, each method of the second kind [where $c(H, E)$ may not be independent of predicates that occur in the language but not in H or E] coincides with some method of the first kind.... Thus, if there is any adequate method of the second kind ..., there is also some adequate method of the first kind" (1963a, p. 781). But we have seen above that methods of the first kind are inadequate. I leave it to the reader to evaluate the force of this objection.

19. The existence of O_T assumes that the observational consequences of T are finitely axiomatizable. When this assumption fails, the second condition can be stated only if conditional probability functions are defined for pairs of sets of sentences. It can be argued that for many interesting theories, O_T is a tautology because no nontrivial observational consequences are derivable without the help of theoretical initial/boundary conditions. In this case the second condition reduces to the first.

20. Carnap might have been taken as having endorsed the possibility of such a utopian scheme through his requirement of *completeness*. But that requirement applies only to the observational language, as is made clear by the following fomulation taken from "On the Application of Inductive Logic": "Every qualitative property or relation of the individuals, that is, every respect in which the positions of the universe may be found to differ by direct observation, must be expressible in L" (Carnap 1947, p. 138). And Carnap's views on the incommensurability of different theoretical frameworks would seem to entail that a universal language for all of science—past, present, and future—is impossible; see "Truth and Confirmation" (1949).

21. K implies that the atoms A and B decay independently; that in the decay process each may emit exactly one of three particles, an α particle, an e^-, or an e^+; that the objective probabilities of these three decay modes are respectively .7, .2, and .1; and that an annihilation event occurs just in case one atom emits an e^- and the other an e^+. T_1 asserts that A emits an e^-, T_2 that B emits an e^-, and E that an annihilation event occurs. The device reported in note 6 can also be used to construct other examples of this sort.

22. I have suppressed the background K to simplify the notation.

23. Additional historical cases are given a similar Bayesian reconstruction in Howson and Urbach 1989.

24. A point emphasized by John Worrall (1991).

Chapter 4

1. The μ in de Finetti's representation theorem (4.2) is the prior distribution over the limiting relative frequency of P's among the $a_1, a_2, \ldots$. That this limit exists with probability 1 follows from some deep results of Birkhoff and von Neumann on stationary process, since an exchangeable process in necessarily stationary (see von Plato 1982). Making these ideas precise requires some of the apparatus not developed until chapter 6 below. For present purposes, the interpretation of μ is not relevant.

For results using concepts of partial exchangeability, see Diaconis and Freedman 1980. For results using ergodicity, see von Plato (1982).

2. For simplicity I have suppressed reference to the background knowledge.

3. But see section 7 for a discussion of just how weak this notion of instance induction is.

4. For another response to Gillies's form of the Popper and Miller argument, see Dunn and Hellman 1986. Yet another move the inductivist can make is to use a different measure of inductive support (see Redhead 1985). Popper and Miller's responses to criticism are to be found in their 1987 paper. See also Howson 1990b.

5. This apt phrase of 'content cutting' is due to Ken Gemes.

6. But see chapter 6, where cases of the nonobjectivity of likelihoods are discussed.

7. Here I am indebted to Philip Kitcher. However, we differ on what philosophy of science can be expected to provide by way of a solution (see chapter 8).

8. The definitions of 'grue' given in the literature are confusing. Goodman defines 'grue' so that "it applies to all things examined before t just in case they are green but to other things just in case they are blue" (1983, p. 74). In the preface to the fourth edition of *Fact, Fiction, and Forecast* Hilary Putnam says that an object is grue "if it is either observed before a certain date and is green, or it is not observed before that date and is blue" (Goodman 1983, p. vii). To get the effect of a shift in color after the magic date that most readers expect from grue, we can employ two-place predicates: Ext, meaning that x is an emerald at t, and Gxt, meaning that x is grue at t, that is, that either t is no later than 2000 and x is green or else t is later than 2000 and x is blue. The hypothesis that all emeralds are grue then reads, $(\forall x)(\forall t)(Ext \to Gxt)$. Alternatively, we can consider a temporally ordered series of objects a_1, a_2, ... and take the grue hypothesis to be $(\forall i)(Ea_i \to Ga_i)$, where Ga_i means that $i \leqslant 2000$ and a_i is green or else $i > 2000$ and a_i is blue. The latter alternative is followed here.

9. Goodman might be read as making a stronger demand on genuine confirmation, namely, that it be homogeneous. One way to unpack the notion of homogeneity is through Hempel's special consequence condition: if E confirms H, then E confirms any logical consequence of H. But as argued in chapter 3, this condition is in general unacceptable if confirmation is interpreted in terms of incremental boosts in probability. A more attractive explication of homogeneity is in terms of exchangeability of the instances of H. But to see Goodmanization whenever there is a failure of homogeneity in this sense is to see Goodmanization throughout science (see the discussion below for more about the connection between exchangeability and Goodman's problem). Ken Gemes (1990b) defines a notion of "real confirmation" requiring that for E to really confirm H, it must incrementally confirm each "content part" of H. Gemes attempts to provide a syntactical definition of the notion of content part that does justice to the intuition that not every logical consequence of H is a genuine content part. To leave aside technical problems with Gemes's definition, my main concern is that his notion of real confirmation is too demanding to be of much use in real scientific examples. For instance, none of the classical tests of Einstein's general theory of relativity or any of the more recent tests I am aware of can plausibly be said to confirm every content part of the theory, and this even though these tests are commonly held to provide good confirmation of the theory. Of course, one could say that individually or collectively these pieces of evidence do confirm every content subpart of some significant part of Einstein's theory. But by the same token, green emeralds confirm every subcontent part of some significant part of 'All emeralds are grue'. In sum, I am unconvinced that there is any workable and useful notion of homogeneous confirmation to be had. And I am convinced that the tools already developed are adequate to capture the valid lessons to be drawn from Goodman's examples.

10. The reason for the qualification 'future-moving' will become evident below.

11. A few years later Carnap abandoned the notion that there is a unique correct confirmation function (e.g., c^*), and in his later years he tended more and more toward personalism.

However, he remained a tempered personalist, for he thought that inductive common sense supplied constraints over and above those of the probability calculus (see Carnap 1980). The problem, of course, is that Carnap's inductive common sense did not agree with the inductive common sense of others. Did he think that there is some overriding *inductive common sense* to which, by some special faculty, he had access? I discuss below how a Carnap aligned with Bayesian personalism should have responded to Goodman's treatment of induction.

12. One of the earliest expressions of this point is to be found in Teller 1969.

13. The word 'grue' was introduced only at a later date.

14. Carnap's papers are part of the Archives for Scientific Philosophy, University of Pittsburgh. The quotation is from document no. 084-19-34, dated January 27, 1946. Quoted by permission of the University of Pittsburgh and C. G. Hempel. All rights reserved.

15. Carnap refereed the manuscript of *Fact, Fiction, and Forecast* for Harvard University Press. He recommended it for publication. Carnap's letter to the press is preserved as document no. 084-19-02 in the Archives for Scientific Philosophy, University of Pittsburgh.

Chapter 5

1. The background knowledge K is suppressed here for the sake of simplicity.

2. The present account skates over some of the historical details. For a full account, see Earman and Glymour 1991.

3. Temporal subscripts on the Pr function will be dropped whenever no confusion will result.

4. The doubts are discussed in detail in chapter 6 below.

5. $\Pr(T \,\&\, \neg E/\neg E) = \Pr(\neg E/\neg E) \times \Pr(T/\neg E) = \Pr(T/\neg E)$ (by (CP3) and (CP2) of appendix 1 of chapter 2). Thus $\Pr(T/\neg E) = 0$ when $T \models E$. When in addition $\Pr(E) = 1$, $\Pr(T/E) = \Pr(T \,\&\, E)/\Pr(E) = \Pr(T)$. So under the stated conditions, $\hat{C}(T, E) = \Pr(T)$.

6. See van Fraassen (1988) who mentions this line without advocating it.

7. See Earman and Glymour 1991 for details and references.

8. Similar problems arise if confirmatory power is measured in other ways, e.g., by Gaifman's (1985) $(1 - \Pr(T))/(1 - \Pr(T/E))$.

9. Here I am using the terminology of Eells 1985. The approach sketched here can also be found in Skyrms 1983.

10. That this was a genuine piece of learning for Einstein is indicated by the fact that he spent about a week of hectic calculation to derive the prediction of perihelion advance from GTR (see Earman and Glymour 1991). And when he found his prediction matched the observed anomaly, he suffered heart palpitations (see Pais 1982, p. 253).

11. Horwich (1982) and Howson (1984, 1985) both advocate that for old evidence, incremental support should be measured in terms of counterfactual degrees of belief. It is not clear, however, at which version of the old-evidence problem their constructions are aimed. If they are aimed at the problem of *old new evidence*—i.e., T was formulated before the discovery of E but it is now later and $\Pr(E) = 1$—then the constructions amount to much the same thing as recommended by Skyrms (1983) and Eells (1985). But if the constructions are aimed at the problem of *new old evidence*—i.e., E was known before the formulation of T and it is now the time of the formulation of T (or barely after it)—or the problem of *old old evidence*—i.e., E was known before the formulation of T and it is now some time subsequent to the formulation—then they are subject to the difficulties discussed in the present section and also in section 7 below.

12. For present purposes, nested iterations of '⊢' will be ignored.

13. For various opinions on this matter, see Howson 1984 and Worrall 1978, 1989. See also the discussion in section 8 of chapter 4.

14. See Hermann 1968, p. 32.

15. See R. Miller 1987, chap. 7, for similar sentiments.

16. For details, see Roseveare 1982 and Earman and Glymour 1991.

17. The problem posed by new theories for orthodox Bayesianism is touched upon by Teller (1975, pp. 173–174). It is discussed explicitly by Chihara (1987). One of the few references in the statistics literature I know of is Leamer 1978, where the problem is discussed under the rubric of "data instigated" hypotheses.

18. See Leamer 1978 for a discussion of the prohibition against doubly counting evidence.

19. Some further remarks about Howson's $K - \{E\}$ are in order. Sometimes he seems to think of K as a discrete set of elements from which E can be plucked. But normally K is treated as being closed under logical implication. (If (LO1) fails, this is an unrealistic assumption, of course.) In this case we might try to get at the relevant sense of $K - \{E\}$ along the following lines. Call K' an intermediary between K and E just in case K' is stated in the vocabulary of K, $K \models K'$, and $K' \not\models E$. If there is a unique strongest intermediary, take this to be $K - \{E\}$. There seems to be no guarantee, however, that there will be a unique strongest intermediary. See also Chihara 1987, pp. 553–554.

Chapter 6

1. Here I am indebted to an unpublished lecture of Peter Railton's, delivered at the Pittsburgh Center for the Philosophy of Science in spring 1990.

2. On this matter, see the discussion in chapters 5 and 8.

3. See Jaynes 1957, 1968. For a critical assessment of the maximum entropy principle, see Seidenfeld 1979, 1986.

4. This is a point that has been repeatedly stressed by Wes Salmon. The Bayesian must deny that these plausibility arguments represent a logically prior form of reasoning that delivers plausibility assessments (see chapter 2).

5. Moving to a rule that determines Pr values not just on a partition but on the entire probability space would be a big step toward apriorism.

6. I have borrowed this example from Wes Salmon.

7. For a discussion of this form of tempered personalism, see Shimony 1970. Note, however, that the sort of criticism given in chapter 8 of Lehrer and Wagner's (1981) proposal for a rule that forms a consensus by means of a weighted aggregation of opinions also applies to this form of tempered personalism.

8. This move falls in with the definitional solution mentioned below in section 9 and discussed critically in chapter 8.

9. As explained in appendix 2 to chapter 2, the strong law of large numbers relies on the countable additivity of the measure on the space of infinite sequences of coin flips. But countable additivity of the degree of belief function Pr is not assumed here. If one wanted to eschew countable additivity altogether, one could fall back on the form of the weak law of large numbers that says that as the number of flips goes to infinity, the probability that the observed frequency of heads differs from the true chance value by any $\varepsilon > 0$ goes to zero. The reader is invited to try to apply this fact to derive a conclusion about merger of opinion.

10. A *random variable* is a map $X: \Omega \to \mathbb{R}$ such that for every Borel set $A \subseteq \mathbb{R}$, $\{\omega \in \Omega: x(\omega) \in A\}$ is in $\mathscr{F}$. Recall that a *field* is a collection of subsets of Ω closed under complementation and finite unions, and that a σ field is closed under countable unions. See appendix 2 of chapter 2 for more details.

11. $E(X)$ is the expectation value of the rv X. $E(X/\mathscr{F}_n)$ stands for the conditional expectation value of X on $\mathscr{F}_n$. If $\mathscr{F}_n$ is the σ field generated by the partition $\{B_j\}$ of Ω, then $E(X/\mathscr{F}_n)$ is defined as the rv such that $E(X/B_j)$ has constant value on each set of the partition.

12. For a different derivation of results similar to those in the theorem, see Schervish and Seidenfeld 1990.

13. The difference between the strong and the weak forms of merger of opinion lies in the order of the quantifiers. For an evidence matrix $\Phi = \{\varphi_i\}$, define the conditional distance measure on probability functions as

$$\text{dist}_{\Phi,n,w}(\text{Pr} - \text{Pr}') \equiv \sup_\psi |\text{Pr}(\psi / \underset{i \leqslant n}{\&} \varphi_i^w) - \text{Pr}'(\psi / \underset{i \leqslant n}{\&} \varphi_i^w)|.$$

Then part (2) of the theorem proves weak merger: for any separating Φ, for a.e. $w \in \text{Mod}_{\mathscr{L}}$, any $\varepsilon > 0$, and for any equally dogmatic Pr and Pr', there exists an N such that for any $m \geqslant N$, $\text{dist}_{\Phi,m,n}(\text{Pr} - \text{Pr}') < \varepsilon$. Strong merger requires that ... there exists an N such that for any equally dogmatic Pr and Pr'....

14. See, however, the discussion below in section 7. An *atomic observation sentence* is a formula of the form $Pi_1 i_2 \ldots i_k$ or $f(i_1, i_2, \ldots, i_k)$ where 'P' and 'f' are respectively a k-ary observation predicate and a k-ary function symbol and the i's are numerals.

15. It is conceivable that the opinions of Bayesian agents regarding a hypothesis could merge in the sense that the absolute difference of the posterior probabilities of any two agents goes to 0 without there being a convergence to certainty on the hypothesis. This raises the question of what conditions need to be assumed in order to prove that merger implies convergence to certainty. As a simple illustration, suppose that H_1 and H_2 are treated as mutually exclusive and exhaustive. And suppose that the class of probability functions $\{\text{Pr}\}$ is *rich enough* that it contains Pr_1 and Pr_2, which differ on the prior odds but agree on the posterior odds for any evidence from the evidence matrix $\Phi = \{\varphi_i\}$. That is,

$$c_1 \equiv \text{Pr}_1(H_1)/\text{Pr}_1(H_2) \neq \text{Pr}_2(H_1)/\text{Pr}_2(H_2) \equiv c_2$$

but

$$R_n \equiv \text{Pr}_1\left(\underset{i \leqslant n}{\&} \pm \varphi_i/H_1\right) \bigg/ \text{Pr}_1\left(\underset{i \leqslant n}{\&} \pm \varphi_i/H_2\right) = \text{Pr}_2\left(\underset{i \leqslant n}{\&} \pm \varphi_i/H_1\right) \bigg/ \text{Pr}_2\left(\underset{i \leqslant n}{\&} \pm \varphi_i/H_2\right)$$

for all n. Now suppose that merger of posterior opinion takes place in the sense that there is an r such that $\text{Pr}_{1,2}(H_1/\underset{i \leqslant n}{\&} \pm \varphi_i) \to r$ as $n \to \infty$. (Note that this is stronger than requiring merger, in the sense that the absolute value of the difference of the posterior probabilities goes to 0, which would allow the opinions to oscillate without settling down to a definite value.) Then r must be 1 or 0. For

$$\text{Pr}_1\left(H_1/\underset{i \leqslant n}{\&} \pm \varphi_i\right) - \text{Pr}_2\left(H_1/\underset{i \leqslant n}{\&} \pm \varphi_i\right) = (c_1 - c_2)/(c_1 + c_2 + R_n + c_1 c_2/R_n).$$

Since this quantity goes to 0 as the limit r is approached, R_n must go either to 0 or to ∞, which imply respectively that the posterior probability of H_1 goes to 0 or 1. For details and a generalization to a countable number of hypotheses, see Hawthorne 1988. Under what other conditions does merger of opinion imply convergence to certainty?

16. I implicitly assumed in section 2 and will explicitly assume in this section that the observational/theoretical distinction can be drawn linguistically by means of a bifurcation of the empirical vocabulary of the language into observational and theoretical terms, the former of which denote directly observable properties or relations. An observation sentence is then

any sentence (not necessarily atomic) all of whose empirical terms are observational. As is well known, this is not the best way to think about the observational/theoretical distinction. Moreover, the further implicit assumption of this section—that the distinction can be drawn in a theory-neutral manner—is also suspect. The justification for using these assumptions here is that they make for a tractable discussion of the merger-of-opinion results. It remains to rework the results when these suspect assumptions are relaxed.

17. Suppose for simplicity that the observational consequences of theory T are finitely axiomatizable and thus representable by an observation sentence O_T. Antirealists have some-times tried to draw comfort from the fact that the probability calculus implies that one's degree of belief in O_T must be greater than or equal to one's degree of belief in T. But from the Bayesian perspective, it is not a question of whether to "prefer" or "choose" T over O_T but rather whether degrees of belief in T can be justified and in particular whether we are ever justified in giving high degrees of belief to interesting theories. See Dorling 1990.

18. Some antirealists, such as van Fraassen (1980), are committed to inductivism with respect to observables, since they hold that we can have good reason to believe that a theory saves all the phenomena.

19. Separation is a necessary condition for the convergence-to-certainty results. Suppose that the evidence matrix $\Phi = \{\varphi_i\}$ is such that for all sentences ψ, $\Pr(\psi/\&_{i \leqslant n} \varphi_i^w) \to [\psi](w)$ for all w in a set of models $K \subseteq \mathrm{Mod}_{\mathscr{L}}$ of measure 1. If Φ didn't separate K, we could choose distinct w_1, $w_2 \in K$ and a sentence π such that $w_1 \in \mathrm{mod}(\pi)$ and $w_2 \in \mathrm{mod}(\neg\pi)$. Then $\Pr(\pi/\&_{i \leqslant n} \varphi_i^{w_1}) \to 1$ and $\Pr(\neg\pi/\&_{i \leqslant n} \varphi_i^{w_2}) \to 0$. This is a contradiction, since if Φ does not separate K, then $\varphi_i^{w_1} = \varphi_i^{w_2}$ for all i.

20. Take the σ-algebra $\mathscr{H} \subseteq \mathscr{F}$ generated by $\mathrm{mod}(T_i)$ and $\mathrm{mod}(\varphi)$, where φ is an observation sentence. The martingale convergence results can then be applied to $(\mathrm{Mod}_{\mathscr{L}}, \mathscr{H}, \mathscr{P}\imath^*)$, where $\mathscr{P}\imath^*$ is the restriction of $\mathscr{P}\imath$ to $\mathscr{H}$, and then transferred down to $\mathscr{L}$.

21. Suppose that wod in the sense of $\widetilde{\mathrm{Mod}}_{\mathscr{L}}$ holds for T_1 and T_2. Consider the collections $\mathring{T}_1$ and $\mathring{T}_2$ of observational consequences of T_1 and T_2 respectively. If these collections were compatible, there would be a common model $\mathring{w}$ (in the observational sublanguage). This common model $\mathring{w}$ need not be extendible to models of T_1 and T_2, but there will exist models w_1 of T_1 and w_2 of T_2 whose reducts to the observational sublanguage will be elementarily equivalent to $\mathring{w}$. This contradicts the assumption of wod. Since $\mathring{T}_1$ and $\mathring{T}_2$ are incompatible, there will be a sentence ψ such that $\mathring{T}_1 \cup \mathring{T}_2 \models \psi \,\&\, \neg\psi$. By compactness, there are finite subsets $\mathring{T} \subset \mathring{T}_1$ and $\mathring{T}_2 \subset \mathring{T}_2$ such that $\mathring{T}_1 \cup \mathring{T}_2 \models \psi \,\&\, \neg\psi$. Let O_1 be the conjunc-tion of the sentences in $\mathring{T}_1$ and let O_2 be the conjunction of the sentences in $\mathring{T}_2$. Then $O_1 \models \neg O_2$, so T_1 and T_2 have respectively the incompatible observational consequences O_1 and O_2. I am indebted to R. A. Rynasiewicz for this demonstration.

22. Let 'P' be an observational predicate, and let the axioms of T_1 consist of $(\exists i)\neg Pi$ and all sentences Pn, where n is a member of a certain set of integers. Let the axioms of T_2 consist of $(\exists i)\neg Pi$ and all sentences Pm, where m is a member of the complement set. T_1 and T_2 are wod, relative to $\mathrm{Mod}_{\mathscr{L}}$. They would also be sod relative to $\widetilde{\mathrm{Mod}}_{\mathscr{L}}$ if the set of integers in ques-tion were definable in $\mathscr{L}$ so that T_1 would entail $(\exists i)(\neg Qi \,\&\, \neg Pi)$ while T_2 would entail $(\forall i)(\neg Qi \to Pi)$ for an arithmetic 'Q'. But there are many sets of integers not definable in $\mathscr{L}$. It is doubtful, however, whether T_1 and T_2 deserve to be called theories, since they are not recursively axiomatizable. I thank R. A. Rynasiewicz for helping me to clarify these matters.

23. However, we will see in chapter 9 that the Bayesian convergence-to-certainty theorem commits the Bayesian to probabilistic certainty that a multiply quantified observational hypothesis is equivalent to a truth-functional compound of hypotheses each of which is either finitely verifiable or falsifiable.

24. As a concrete case of what I find to be an interesting example of underdetermination, I mention flat and nonflat versions of Newtonian gravitational theory that predict the same particle orbits (see Anderson 1967). For an argument against underdetermination in an important incident in the development of the old quantum theory, see Norton 1990.

25. Recall the discussion in chapter 4 of the past-reaching sense of projectability. Thus Hume's problem of induction and Goodman's so-called new problem can both be seen as a species of the problem of underdetermination.

26. Hans Reichenbach consistently maintained the latter criterion, which is still defended by Salmon (1979). In *Experience and Prediction* Reichenbach wrote, "The probability theory of meaning ... allows us to maintain propositions as meaningful which concern facts outside the domain of the immediately verifiable facts; it allows us to pass beyond the domain of given facts. This *overreaching* character of probability inferences is the basic method of the knowledge of nature" (1961, p. 127). My point here is that on the Bayesian personalist interpretation of probability, this overreaching character threatens to extend into areas the positivists and logical empiricists wanted to exclude. Of course, Reichenbach maintained a relative-frequency interpretation of probability. Whether this interpretation is defensible and, if so, whether it provides a resolution of the problems raised above remain to be seen. In his later years Carnap moved toward a subjectivist criterion of meaningfulness: "I regard as meaningful for me whatever I can, in principle, confirm subjectively" (1963c, p. 882). This approach threatens to open the floodgates. However, Carnap also espoused a form of physicalism that he took to imply, "Whatever is subjectively confirmable, is also intersubjectively confirmable" (1963c, p. 883).

27. It might be objected that this is an unfair example, since the weather is unpredictable. I suspect, however, that this complaint amounts to the claim that, say, my degree of belief of 1/2 that it will rain in Pittsburgh tomorrow has no natural interpretation as an estimate of frequencies. If true, this would undermine the present attempt to interpret the objectivity of degrees of belief. In any case, the point of the example stands: the ability to tolerate unanticipated circumstances may be as advantageous in evolutionary terms as the ability to anticipate. In criticism of the evolutionary explanation I have also heard it said that overestimating the probabilities of dangers may be evolutionary advantageous. I am not persuaded of this point, since, for example, the behavior of an animal that shies at every apparent danger can be explained by assuming accurate estimates of the likelihood of the dangers and large negative utilities for having to face the dangers.

28. Another difficulty here stems from the utopian assumption that the agent can list a complete partition of theories (see chapter 7).

Chapter 7

1. Or perhaps the hypothesis or theory is falsifiable, but only with the help of auxiliary assumptions that themselves are not finitely verifiable and that therefore are known at best only with a high probability.

2. This passage does not appear in the third edition of Giere's book. My purpose here is not to pick on my friend Ron for a view that he may no longer hold but rather to probe what is a widely shared prejudice against eliminative induction.

3. However, we will see in chapter 9 that the Bayesian convergence-to-certainty theorem commits the Bayesian to probabilistic certainty that a multiply quantified observational hypothesis is equivalent to a truth-functional compound of hypotheses each of which is finitely verifiable or falsifiable.

4. If antirealists are to be believed, it can never be had, because theory is underdetermined by observational evidence precisely in the sense that there exist theories not observationally distinguishable from their rivals. For some remarks on this matter, see chapter 6 and section 4 below.

5. In *A Treatise on Probability* Keynes wrote, "The conception of having *some* reason, though not a conclusive one, for certain beliefs, arising out of direct inspection, may prove important to the theory of epistemology. The old metaphysics has been greatly hindered by reason of its having always demanded demonstrative certainty.... When we allow that probable knowledge is, nevertheless, real, a new method of argument can be introduced into metaphysical discussions" (1962, pp. 239–240). It is this "new method of argument" that Russell attempted to clarify in *Human Knowledge: Its Scope and Limits* (1948).

6. I have changed Giere's notation to conform to mine and have included the "initial conditions" in K. Again, my purpose here is not to pick on Giere but to criticize what I take to be a vain hope about theory testing. The hope goes back to Keynes and Russell and is still widely shared.

7. I have changed Jeffreys's notation to conform to mine.

8. There are still problems aplenty here (see chapter 6).

9. In fairness to my revered colleague Wes Salmon, I should point out that his proposal was made in the context of trying to understand Kuhn's account of theory choice. However, I would note that Bayesians should not get sucked into playing the game of choosing among theories (see chapter 8).

10. See Earman and Glymour 1980b for an account of the British eclipse expeditions and their role in the early reception of GTR.

11. See Norton 1989 for a detailed account of how Einstein discovered and justified GTR to himself. According to Norton, Einstein's reasoning follows eliminativist lines.

12. For an analysis of Einstein's derivation of the perihelion shift, see Earman and Glymour 1991.

13. See Earman and Glymour 1980a for a review of the early red-shift tests.

14. For more details of Eddington's analysis, see Earman and Glymour 1980b.

15. For a comprehensive review of these and other members of the zoo of gravitational theories, see Will 1981.

16. Other relevant references describing the program include Will 1971a, 1971b, 1972, 1974a, 1974b, 1981, 1984; Ni 1972, 1973; Thorne, Lee, and Lightman 1973.

17. For a definition of this and other terms relevant to Thorne and Will's program, see Thorne, Lee, and Lightman 1973.

18. For example, to explain the influence of gravity on light, either Maxwell's theory or the photon theory may be used.

19. Will defines a test body as one "that has negligible self-gravitational energy (as estimated in Newtonian theory) and is small enough in size so that its coupling to the inhomogeneities of external fields can be ignored" (1984, p. 351).

20. This is a reference frame that falls along one of the geodesics of **g** and is small enough that the inhomogeneities of the gravitational field can be ignored.

21. In recent years there has been some speculation about a "fifth force." The existence of such a force would undermine the weak principle of equivalence, since it is supposed to moderate the gravitational attraction of bodies in a manner that depends upon the chemical composition of the bodies. The evidence is now deemed to be overwhelmingly against such a force. It

would be interesting to see how this judgment can be rationalized in terms of the Bayesian-eliminative scheme proposed here.

22. This figure is adapted from Will 1972, 1974b.

23. Apologies may be due to Sherlock Holmes for using him as a foil. Although many passages suggest that he believed in a crude form of eliminative induction, other passages indicate a more sophisticated view. (Here I am indebted to Prof. Soschichi Uchii.) In *The Sign of Four* Holmes speaks of the "balance of probability," indicating that his method involves probabilistic elements. And even more intriguing, in *The Hound of the Baskervilles* he speaks of "the scientific use of the imagination." With a little charity we can read this as the ability to articulate all of the relevant possibilities. Thus Holmes's success may be seen as the result of carrying out the version of eliminative induction discussed above.

Chapter 8

1. *Structure*, p. 199. This already serves to separate Kuhn from Quine's epistemology and all of its disastrous consequences.

2. For Bayesianism, the emergence of a sharp consensus (in the degree-of-belief sense) from widely differing prior probabilities is at the heart of scientific objectivity (see chapter 6 for a discussion of objectivity and rationality within the Bayesian framework).

3. See Friedman 1983 and Earman 1989 for details.

4. While the use of this language may not be the appropriate device for understanding all of the historical disputes, it does illuminate those having to do with the long-running debate between the absolute versus relational conceptions of space and time (see my 1989 book). There may be some sense in which Newton's theory, translated into space-time language, is no longer Newton's theory. But I deny that there is any interesting sense in which this translation fails to accurately capture the physical content of Newton's theory. Nevertheless, I acknowledge that the adoption of the new language influences probability assignments. I will return to this point in sections 3 and 6.

5. Since I have never been able to understand exactly what is at issue here, I don't know whether I should demur or not.

6. My response to incommensurability and relativism has been tactical. A more radical response is given by Kelly and Glymour (1990), who argue that there are inductive methodologies powerful enough to take into account shifting frameworks, paradigms, and whatnot.

7. At some junctures Giere (1988) distinguishes between choosing to pursue a theory and accepting the theory as true; at other places, however, it is hard to tell what he has in mind. The issues discussed here have played themselves out before in philosophy of science (see, for example, Bar-Hillel 1968, Carnap 1968, and Kyburg 1968).

8. It might be complained that successful puzzle solving in normal science requires a commitment to or an embrace of a theory that goes beyond a practical decision to allocate a portion of one's research time to the theory. This complaint may or may not be correct as a psychological account of the motivation of individual scientists. I think that it is true for some scientists but false for others.

I should also note that there is a sentiment in favor of theory acceptance on the grounds that it is needed to produce scientific explanations (see, for example, Salmon 1968). I must confess that I find this sentiment nearly incomprehensible. Without attending to the truth of a theory *T*, we can check the various conditions of Hempel and Oppenheim's account (or your favorite alternative account) of scientific explanation to make sure that *T* is a potential explanation of the phenomenon under scrutiny, that is, that *T*, if true, would be an explanation. The notion that we should then accept T, or some other rival potential explainer, to turn the potential into an actuality strikes me as being akin to thinking that wishing it were so makes it so. If on the basis of the available evidence the probability of T and each of its rivals is less than 1, as it almost always is, then that's the way things are, and wishing it were otherwise is futile.

Finally, I want to comment on the notion of accepting a theory in the sense of "acting on it." Thus scientists may be said to act on Newton's theory when they use it to calculate the orbit of a lunar rocket probe. But the very same scientists, without any change of cognitive attitudes, may act on Einstein's GTR rather than Newton's theory in calculating planetary perihelia. What we again have are practical decisions whose different outcomes turn on the different utilities assigned to accuracy of prediction and ease of calculation in the two contexts.

9. Which form of revolution was occasioned by the introduction of Einstein's GTR? Arguments can be made in favor of both the weak and strong forms.

10. For details, see Earman and Glymour 1991.

11. Lehrer and Wagner prove various results about the conditions under which iterated weighted aggregation produces a consensual probability; the details are not relevant here.

12. Donald Gillies (1990) has argued for what he calls "intersubjective probabilities" on the grounds that Dutch book can be made against a community whose members have differing degrees of belief. This argument proves too much, since it would entail the unhealthy consequence that the members of a scientific community are not allowed to disagree. The way to escape the conclusion is to deny that a scientific community acts as a corporate or economic body in discharging its basic role of discovering and testing scientific hypotheses. In addition, Gillies's communal Dutch-book construction is not amenable to the more pristine interpretation (recommended in chapter 2) that relegates Dutch bookies to the role of window dressing.

13. That is, unless it could be shown that the class of belief functions has been narrowed for evolutionary reasons that enhance the reliability of the degrees of belief, an option I found unattractive in chapter 6.

14. As I learned from David Hull's lecture "Why Scientists Behave Scientifically," delivered in the Pittsburgh Series in the Philosophy of Science, September 1990. Philip Kitcher pointed out to me that Merton himself offered an explanation along these lines (see Merton 1973).

Chapter 9

1. This is a slight modification of the setup used in chapter 6.

2. See chapter 14 of Rogers 1987.

3. A "positive instance" $I(n)$ of H_D will have the form Pn & $\neg Ra_n$ or $\neg Pn$ & Ra_n. Since evidence statements are assumed to be true and since if Pn is true (false) in any $w \in \text{Mod}_{\mathscr{S}}$, it is true (false) in every $w \in \text{Mod}_{\mathscr{S}}$, it follows that for positive instances we can set $\Pr(Pn$ & $\neg Ra_n) = \Pr(\neg Ra_n)$ and $\Pr(\neg Pn$ & $Ra_n) = \Pr(Ra_n)$. Thus $\Pr(H_D/\&_{i \leqslant n} I(n)) = \Pr(H_D/\&_{i \leqslant n} I'(n))$, where $I'(n)$ is $\neg Ra_n$ or Ra_n according as $I(n)$ is Pn & $\neg Ra_n$ or $\neg Pn$ & Ra_n. As a result, we can continue to take evidence sequences to be of the form Ra_1 & $\neg Ra_2$ & ....

4. From note 3 we know that

$$\Pr(H_D/\underset{i \leqslant n}{\&} I(n)) = \Pr(H_D/\underset{i \leqslant n}{\&} I'(n))$$

$$= \frac{\Pr(H_D)}{\Pr(I'(1)) \times \Pr(I'(2)/I'(1)) \times \cdots \times \Pr(I'(n)/\&_{i \leqslant n-1} I'(i))}.$$

The results of chapter 4 can now be applied to conclude that if H_D is true and $\Pr(H_D) > 0$, the instance confirmation goes to 1 as the number of positive instances goes to infinity and that if countable additivity holds, the probability of H_D itself goes to 1 in the limit. Recall that in this context, countable additivity means that

$$\Pr((\forall i)\eta(i)) = \lim_{n\to\infty} \Pr(\underset{j \leqslant n}{\&}\, \eta(j)).$$

5. For results relevant to the PAC (or probable approximate convergence) model of learning, see Blumer, Ehrenfeucht, Haussler, and Warmuth 1987 and Rivest et al. 1989.

6. For more careful and detailed formulations of formal-learning problems, see Osherson, Stob, and Weinstein 1986; Osherson and Weinstein 1989a, 1989b; and Kelly and Glymour 1989. The paradigm adopted here of truth detection in the limit is from Kelly and Glymour 1989.

7. The domain of any $w \in \mathrm{Mod}_{\mathscr{L}}$ is the disjoint union of two parts D_1 and D_2, each of which is countably infinite. Each element of D_1 is named by a numeral, each element of D_2 is named by an a_i, and n-ary arithmetic and empirical predicates are interpreted subsets of the n-fold Cartesian products $D_1 \times D_1 \times \cdots \times D_1$ and $D_2 \times D_2 \times \cdots \times D_2$ respectively.

8. In a different setting, Osherson, Stob, and Weinstein (1988) show that effective Bayesian learners must pay a price.

9. The term 'verifutable' introduced above is due to Kelly and Glymour (1989). The example used here is due to Kelly (1990).

10. To my knowledge, Kevin Kelly was the first to use formal-learning theory to show how Bayesianism contains a concealed dogmatism. I am grateful to him for sharing his work with me prior to its publication.

11. If, say, $K = \mathrm{mod}((\exists i)\,\neg Pa_i)$, then K is not compact in the evidence, since every finite subset of $\{Pa_1, Pa_2, \ldots\}$ is satisfiable in K, even though the entire set is not.

12. Compare this to the usual notion of not finitely verifiable/falsifiable obtained by negating the condition that ψ is *finitely verifiable* (respectively, *falsifiable*) relative to K just in case for any $w \in K$ such that $w \in \mathrm{mod}(\psi)$ (respectively, $w \in \mathrm{mod}(\neg\psi)$), there is a finite E such that $w \in \mathrm{mod}(E)$ and such that for any $w' \in K$, if $w' \in \mathrm{mod}(E)$, then $w' \in \mathrm{mod}(\psi)$ (respectively, $w' \in \mathrm{mod}(\neg\psi)$).

13. In the general case where $K = \bigcup_i \mathrm{mod}(\Delta_i)$, the Bayesian agent is probabilistically certain at the outset that the actual world models one of the Δ_i, and that relative to any such Δ_i, ψ is equivalent to a verifutable sentence.

14. For a sampling of recent opinions on the reliabilist conception of knowledge, see Alston 1986, Bonjour 1985, Foley 1985, and Goldman 1986.

15. The formal learner may be unmoved by this line of argument in view of the fact that, short of the infinite limit, the Bayesians have not been able to give any objective significance to the little numbers they assign (see chapter 6).

16. See Earman 1986 and Penrose 1989 for suggestions along this line.

17. This seems to be what Putnam has in mind when he says that somebody proposes the hypothesis.

18. See Carnap's 1963b response to Putnam.

Chapter 10

1. Here Sue seems to be referring to van Fraassen's objection, discussed in section 3 of chapter 5 of *Bayes or Bust?*

References

Abraham, M.

1912. "Zur Theorie der Gravitation." *Physikalische Zeitschrift* 13:1–4.

Aczél, J.

1966. *Lectures on Functional Equations and Their Applications*. New York: Academic Press.

Adams, E.

1961. "On Rational Betting Systems." *Archiv für mathematische Logik und Grundlagenforschung* 6:7–29, 112–128.

Adams, E., and Rosenkrantz, R. D.

1980. "Applying the Jeffrey Decision Model to Rational Betting and Information Acquisition." *Theory and Decision* 12:1–20.

Allais, M.

1953. "Le comportement de l'homme rationel devant le risque: Critiques des postulats et axiomes de l'école americaine." *Econometrika* 21:503–546.

Alston, W.

1986. "Internalism and Externalism in Epistemology." *Philosophical Topics* 14:179–221.

Anderson, J. L.

1967. *Principles of Relativity Physics*. New York: Academic Press.

Aronson, J.

1989. "The Bayesians and the Ravens Paradox." *Noûs* 23:221–240.

Arrow, K.

1971. *Essays in the Theory of Risk-Bearing*. Chicago: Markham Press.

Bar-Hillel, Y.

1968. "On Alleged Rules of Detachment in Inductive Logic." In Lakatos 1968.

Barnard, G. A.

1958. "Thomas Bayes—A Biographical Note." *Biometrika* 45:293–295.

Bayes, T.

1764. "An Essay Towards Solving a Problem in the Doctrine of Chances." *Philosophical Transactions of the Royal Society of London* 53:370–418. Reprinted in facsimile in W. E. Deming, ed., *Facsimiles of Two Papers by Bayes* (Washington, D.C.: U.S. Dept. of Agriculture, 1940). New edition published in *Biometrika* 45 (1958):296–315; all page references are to this edition. Also in E. S. Pearson and M. G. Kendall, eds., *Studies in the History of Statistics and Probability* (London: Charles Griffin, 1970).

Berenstein, C., Kanal, L. N., and Lavine, D.

1986. "Consensus Rules." In L. N. Kanal and J. F. Lemmer, eds., *Uncertainty in Artificial Intelligence* (Amsterdam: North-Holland).

Billingsley, P.

1979. *Probability and Measure*. New York: John Wiley.

Birkhoff, G. D.

1943. "Matter, Electricity, and Gravitation in Flat Space-Time." *Proceedings of the National Academy of Sciences* 29:231–239.

Blumer, A., Ehrenfeucht, A., Haussler, D., and Warmuth, M. K.

1987. "Learnability and the Vapnik-Chervonenkis Dimension." Technical report UCSC-CRL-87-20, University of California of Santa Cruz.

Bonjour, L.

1985. *The Structure of Empirical Knowledge*. Cambridge: Harvard University Press.

Boole, G.

1854. *An Investigation of the Laws of Thought*. New York: Dover, 1954.

Borel, E.

1924. "Apropos of a Treatise on Probability." *Revue Philosophique* 98:321–336. Reprinted in Kyburg and Smokler 1964.

Braginsky, V. B., and Panov, V. I.

1972. "Verification of the Equivalence of Inertial and Gravitational Mass." *Soviet Physics Journal of Theoretical and Experimental Physics* 34:463–466.

Brush, S. G.
1989. "Prediction and Theory Evaluation: The Case of Light Bending." *Science* 246: 1124–1129.
Burks, A. W.
1977. *Chance, Cause, Reason.* Chicago: University of Chicago Press.
Carnap, R.
1947. "On the Application of Inductive Logic." *Philosophy and Phenomenological Research* 8:133–147.
1949. "Truth and Confirmation." In H. Feigl and W. Sellars, eds., *Readings in Philosophical Analysis* (New York: Appleton-Century-Crofts).
1950. *Logical Foundations of Probability.* Chicago: University of Chicago Press.
1952. *The Continuum of Inductive Methods.* Chicago: University of Chicago Press.
1963a. "Intellectual Autobiography." In Schilpp 1963.
1963b. "Hilary Putnam on Degree of Confirmation and Inductive Logic." In Schilpp 1963.
1963c. "Herbert Feigl on Physicalism." In Schilpp 1963.
1968. "On Rules of Acceptance." In Lakatos 1968.
1980. "A Basic System of Inductive Logic, Part 2." In Jeffrey 1980.
Chihara, C.
1987. "Some Problems for Bayesian Confirmation Theory." *British Journal for the Philosophy of Science* 38:551–560.
1988. "An Interchange on the Popper-Miller Argument." *Philosophical Studies* 54:1–8.
Christensen, D.
1983. "Glymour on Evidence and Relevance." *Philosophy of Science* 50:471–481.
1990. "The Irrelevance of Bootstrapping." *Philosophy of Science* 57:644–662.
Cox, R. T.
1946. "Probability, Frequency, and Reasonable Belief." *American Journal of Physics* 14: 1–13.
1961. *The Algebra of Probable Inference.* Baltimore: Johns Hopkins University Press.
Craig, W.
1956. "Replacement of Auxiliary Expressions." *Philosophical Review* 65:38–55.
Dale, A. I.
1982. "Bayes or Laplace? An Examination of the Origin and Early Applications of Bayes' Theorem." *Archive for History of Exact Sciences* 27:23–47.
1986. "A Newly Discovered Result of Thomas Bayes'." *Archive for the History of Exact Sciences* 35:101–113.
Dawid, P., and Gillies, D.
1989. "A Bayesian Analysis of Hume's Argument Concerning Miracles." *Philosophical Quarterly* 39:57–65.
De Finetti, B.
1937. "La prévision: Ses lois logiques, ses sources subjectives." *Annales de l'Institut Henri Poincaré* 7:1–68. English translation in Kyburg and Smokler 1964.
1972. *Probability, Induction and Statistics.* New York: John Wiley.
Diaconis, P., and Freedman, D.
1980. "De Finetti's Generalizations of Exchangeability." In Jeffrey 1980.
Diaconis, P., and Zabell, S.
1982. "Updating Subjective Probability." *Journal of the American Statistical Association* 77:822–830.
Dicke, R. H.
1964. "Experimental Relativity." In C. DeWitt and B. DeWitt, eds., *Relativity, Groups, and Topology* (New York: Gordon and Breach).
Doob, J. L.
1941. "Probability as Measure." *Annals of Mathematical Statistics* 12:206–214.
1971. "What Is a Martingale?" *American Mathematical Monthly* 78:451–462.

Dorling, J.
1979. "Bayesian Personalism, the Methodology of Scientific Research Programs, and Duhem's Problem." *Studies in the History and Philosophy of Science* 10:177–187.
1990. "Three Recent Breakthroughs in the Probabilistic/Inductivist Approach to Scientific Inference." Preprint.
Dunn, M., and Hellman, G.
1986. "Dualling: A Critique of an Argument of Popper and Miller." *British Journal for the Philosophy of Science* 37:220–223.
Earman, J.
1986. *A Primer on Determinism*. Dordrecht: D. Reidel.
1989. *World Enough and Space-Time: Absolute versus Relational Theories of Space and Time*. Cambridge: MIT Press.
Earman, J., ed.
1983. *Testing Scientific Theories*. Minnesota Studies in the Philosophy of Science, vol. 10. Minneapolis: University of Minnesota Press.
Earman, J., and Glymour, C.
1980a. "The Gravitational Redshift as a Test of Einstein's General Theory of Relativity: History and Analysis." *Studies in History and Philosophy of Science* 11:175–214.
1980b. "Relativity and the Eclipses: The British Eclipse Expeditions and Their Predecessors." *Historical Studies in the Physical Sciences* 11:49–85.
1991. "Einstein's Explanation of the Motion of Mercury's Perihelion." *Einstein Studies*. Forthcoming.
Eddington, A. S.
1923. *The Mathematical Theory of Relativity*. Cambridge: Cambridge University Press.
Edidin, A.
1983. "Bootstrapping without Bootstraps." In Earman 1983.
1988. "From Relative to Real Confirmation." *Philosophy of Science* 55:265–271.
Edwards, A. W. F.
1978. "Commentary on the Arguments of Thomas Bayes." *Scandinavian Journal of Statistics* 5:116–118.
Edwards, W., Lindman, H., and Savage, L. J.
1963. "Bayesian Statistical Inference for Psychological Research." *Psychological Review* 70:193–242.
Eells, E.
1985. "Problems of Old Evidence." *Pacific Philosophical Quarterly* 66:283–302.
Einstein, A.
1912a. "Lichtgeschwindigkeit and Statik des Gravitätionsfeldes." *Annalen der Physik* 38:355–369.
1912b. "Zur Theorie des statischen Gravitätionsfeldes." *Annalen der Physik* 38:443–458.
Einstein, A., and Grossmann, M.
1913. "Entwurf einer verallgemeinerten Relativitätstheorie und einer Theorie der Gravitätion." *Zeitschrift für Mathematik und Physik* 62:225–261.
Ellsberg, D.
1961. "Risk, Ambiguity, and the Savage Axioms." *Quarterly Journal of Economics* 75:643–669.
Field, H.
1978. "A Note on Jeffrey Conditionalization." *Philosophy of Science* 45:361–367.
Fisher, R. A.
1922. "On the Mathematical Foundations of Theoretical Statistics." *Philosophical Transactions of the Royal Society of London*, series A, 222:309–368.
1956. *Statistical Methods and Scientific Inference*. Edinburgh: Oliver and Boyd.
Fodor, J.
1984. "Observation Reconsidered." *Philosophy of Science* 51:23–41.
Foley, R.
1985. "What's Wrong with Reliabilism?" *Monist* 68:188–202.

Franklin, A.

1986. *The Neglect of Experiment*. Cambridge: Cambridge University Press.

1990. *Experiment, Right or Wrong*. Cambridge: Cambridge University Press.

Franklin, A., and Howson, C.

1984. "Why Do Scientists Prefer to Vary Their Experiments?" *Studies in the History and Philosophy of Science* 15:51–62.

French, S.

1988. "A Green Parrot Is Just as Much a Red Herring as a White Shoe: A Note on Confirmation, Background Knowledge, and the Logico-probabilistic Approach." *British Journal for the Philosophy of Science* 38:531–535.

Friedman, M.

1983. *Foundations of Space-Time Theories*. Princeton: Princeton University Press.

Gaifman, H.

1979. "Subjective Probability, Natural Predicates, and Hempel's Ravens." *Erkenntnis* 14: 105–147.

1985. "On Inductive Support and Some Recent Tricks." *Erkenntnis* 22:5–21.

Gaifman, H., and Snir, M.

1982. "Probabilities over Rich Languages." *Journal of Symbolic Logic* 47:495–548.

Garber, D.

1980. "Field and Jeffrey Conditionalization." *Philosophy of Science* 47:142–145.

1983. "Old Evidence and Logical Omniscience in Bayesian Confirmation Theory." In Earman 1983.

Geisser, S.

1980. "A Predictivistic Primer." In A. Zellner, ed., *Bayesian Analysis in Econometrics and Statistics* (Amsterdam: North-Holland).

Gemes, K.

1990a. "Horwich, Hempel, and Hypothetico-deductivism." *Philosophy of Science* 57:699–702.

1990b. "Content and Confirmation." Ph.D. thesis, Dept. of Philosophy, University of Pittsburgh.

Giere, R.

1984. *Understanding Scientific Reasoning*. 2nd ed. New York: Holt, Rinehart, and Winston. 3rd ed., 1991. Page references are to the second edition.

1988. *Explaining Science*. Chicago: University of Chicago Press.

Gillies, D.

1986. "In Defense of the Popper-Miller Argument." *Philosophy of Science* 53:111–113.

1987. "Was Bayes a Bayesian?" *Historica Mathematica* 14:325–346.

1990. "Intersubjective Probability and Confirmation Theory." *British Journal for the Philosophy of Science*, forthcoming.

Glymour, C.

1980. *Theory and Evidence*. Princeton: Princeton University Press.

1983. "Revisions of Bootstrap Testing." *Philosophy of Science* 50:626–629.

Goldman, A. I.

1986. *Epistemology and Cognition*. Cambridge: Harvard University Press.

Good, I. J.

1950. *Probability and the Weighing of Evidence*. London: Charles Griffin.

1967. "The White Shoe Is a Red Herring." *British Journal for the Philosophy of Science* 17:322.

1977. "Dynamic Probability, Computer Chess, and the Measurement of Knowledge." In E. W. Elcock and D. Mitchie, eds., *Machine Intelligence*, vol. 8 (New York: Halsted Press).

Goodman, N.

1946. "A Query on Confirmation." *Journal of Philosophy* 43:383–385.

1947. "On Infirmities of Confirmation-Theory." *Philosophy and Phenomenological Research* 8:149–151.

1983. *Fact, Fiction, and Forecast.* 4th ed. Cambridge: Harvard University Press.

Grünbaum, A.

1976. "Is Falsifiability the Touchstone of Scientific Rationality? Karl Popper versus Inductivism." In R. S. Cohen, P. K. Feyerabend, and M. W. Wartofsky, eds., *Essays in Memory of Imre Lakatos* (Dordrecht: D. Reidel).

Hacking, I.

1965. *Logic of Statistical Inference.* Cambridge: Cambridge University Press.

1970. "Thomas Bayes." *Dictionary of Scientific Biography,* vol. 1. New York: Charles Scribner's Sons.

1975. *The Emergence of Probability.* Cambridge: Cambridge University Press.

Hawthorne, J.

1988. "A Semantic Theory for Partial Entailment and Inductive Inference." Ph.D. thesis, University of Minnesota.

Hempel, C. G.

1945. "Studies in the Logic of Confirmation." *Mind* 54:1–26, 97–121. Reprinted in Hempel 1965.

1958. "The Theoretician's Dilemma." In H. Feigl, G. Maxwell, and M. Scriven, eds., *Concepts, Theories, and the Mind-Body Problem,* Minnesota Studies in the Philosophy of Science, vol. 2 (Minneapolis: University of Minnesota Press). Reprinted in Hempel 1965.

1965. *Aspects of Scientific Explanation.* New York: Free Press.

Hermann, A., ed.

1968. *Albert Einstein/Arnold Sommerfeld Briefwechsel.* Basel: Schabe and Co.

Hesse, M.

1975. "Bayesian Methods and the Initial Probability of Theories." In Maxwell and Anderson 1975.

Hintikka, J.

1966. "A Two-Dimensional Continuum of Inductive Methods." In Hintikka and Suppes 1966.

Hintikka, J., and Niiniluoto, I.

1980. "An Axiomatic Foundation for the Logic of Inductive Generalization." In Jeffrey 1980.

Hintikka, J., and Suppes, P., eds.

1966. *Aspects of Inductive Logic.* Amsterdam: North-Holland.

Horwich, P.

1982. *Probability and Evidence.* Cambridge: Cambridge University Press.

1983. "Explanations of Irrelevance." In Earman 1983.

Howson, C.

1973. "Must the Logical Probability of Laws Be Zero?" *British Journal for the Philosophy of Science* 24:153–163.

1984. "Bayesianism and Support by Novel Facts." *British Journal for the Philosophy of Science* 35:245–251.

1985. "Some Recent Objections to the Bayesian Theory of Support." *British Journal for the Philosophy of Science* 36:305–309.

1990a. "Subjective Probabilities and Betting Quotients." *Synthese* 81:1–8.

1990b. "Some Further Reflections on the Popper-Miller 'Disproof' of Probabilistic Induction." Preprint.

Howson, C., and Urbach, P.

1989. *Scientific Reasoning: The Bayesian Approach.* La Salle: Open Court.

Hume, D.

1739. *Treatise of Human Nature.* Ed. L. A. Selby-Bigge. Oxford: Oxford University Press, 1965.

1748. *Enquiry into the Human Understanding.* Reprinted from the 1777 edition in L. A. Selby-Bigge and P. H. Nidditch eds., *Enquiries Concerning Human Understanding and Concerning the Principles of Morals* (Oxford: Oxford University Press, 1975).

Jaynes, E. T.
  1957. "Information Theory and Statistical Mechanics, I and II." *Physical Review* 106:
    620–630; 108:171–190.
  1959. *Probability Theory in Science and Engineering.* Dallas: Socony Mobil Oil Co.
  1968. "Prior Probabilities." *I.E.E.E. Transactions on Systems Science and Cybernetics,*
    SSC-4:227–241.
Jeffrey, R. C.
  1970. "Review of Eight Discussion Notes." *Journal of Symbolic Logic* 35:124–127.
  1983a. "Bayesianism with a Human Face." In Earman 1983.
  1983b. *The Logic of Decision.* 2nd ed. New York: McGraw-Hill.
  1984. "The Impossibility of Inductive Probability." *Nature* 310:433.
Jeffrey, R. C., ed.
  1980. *Studies in Inductive Logic and Probability.* Vol. 2. Berkeley: University of California
    Press.
Jeffreys, H.
  1961. *Theory of Probability.* 3rd ed. Oxford: Oxford University Press.
  1973. *Scientific Inference.* 3rd ed. Cambridge: Cambridge University Press.
Johnson, W. E.
  1932. "Probability: The Deductive and Inductive Problems." *Mind* 49:409–423.
Kadane, J. B., Schervish, M. J., and Seidenfeld, T.
  1986. "Statistical Implications of Finitely Additive Probability." In P. Goel and A. Zellner,
    eds., *Bayesian Inference and Decision Techniques* (Amsterdam: Elsevier).
Kadane, J. B., and Winkler, R. L.
  1987. "De Finetti's Methods of Elicitation." In R. Viertl, ed., *Probability and Bayesian
    Statistics* (New York: Plenum Press).
  1988. "Separating Probability Elicitation from Utilities." *Journal of the American Statisti-
    cal Association* 83:357–363.
Kahneman, D., Slovic, P., and Tversky, A., eds.
  1982. *Judgment under Uncertainty: Heuristics and Biases.* Cambridge: Cambridge Univer-
    sity Press.
Kelly, K.
  1990. *The Logic and Complexity of Scientific Discovery.* Preprint.
Kelly, K., and Glymour, C.
  1989. "Convergence to the Truth and Nothing but the Truth." *Philosophy of Science*
    56:185–220.
  1990. "Inductive Inference from Theory Laden Data." Preprint.
Kemeny, J. G.
  1955. "Fair Bets and Inductive Probabilities." *Journal of Symbolic Logic* 20:263–273.
  1963. "Carnap's Theory of Probability and Induction." In Schilpp 1963.
Keynes, J. M.
  1962. *A Treatise on Probability.* New York: Harper and Row.
Klibansky, R., and Mossner, E. C., eds.
  1954. *New Letters of David Hume.* Oxford: Oxford University Press.
Kuhn, T. S.
  1962. *The Structure of Scientific Revolutions.* Chicago: University of Chicago Press.
  1970. *The Structure of Scientific Revolutions.* 2nd ed. Chicago: University of Chicago Press.
  1977. "Objectivity, Value Judgments, and Theory Choice." In *The Essential Tension* (Chi-
    cago: University of Chicago Press).
  1983. "Rationality and Theory Choice." *Journal of Philosophy* 80:563–570.
  1989. "Possible Worlds in History of Science." In S. Allen, ed., *Possible Worlds in Hu-
    manities, Arts, and Sciences* (Berlin: W. de Gruyter).
Kustaanheimo, P., and Nuotio, V. S.
  1967. "Relativity Theories of Gravitation. I: One-Body Problem." Unpublished, Dept. of
    Applied Mathematics, University of Helsinki.

Kyburg, H.
    1968. "The Role of Detachment in Inductive Logic." In Lakatos 1968.
    1974. *The Logical Foundations of Statistical Inference.* Dordrecht: D. Reidel.
Kyburg, H., and Smokler, H., eds.
    1964. *Studies in Subjective Probability.* New York: John Wiley.
Lakatos, I.
    1970. "Falsification and the Methodology of Scientific Research Programs." In I. Lakatos
    and A. Musgrave, eds., *Criticism and the Growth of Knowledge* (Cambridge: Cambridge
    University Press).
Lakatos, I., ed.
    1968. *The Problem of Inductive Logic.* Amsterdam: North-Holland.
Lawson, T.
    1985. "The Context of Prediction (and the Paradox of Confirmation)." *British Journal for
    the Philosophy of Science* 36:393–407.
Leamer, E. E.
    1978. *Specification Searches.* New York: John Wiley.
Lehman, R. S.
    1955. "On Confirmation and Rational Betting." *Journal of Symbolic Logic* 20:251–262.
Lehrer, K., and Wagner, C.
    1981. *Rational Consensus in Science and Society.* Dordrecht: D. Reidel.
Levi, I.
    1980. *The Enterprise of Knowledge.* Cambridge: MIT Press.
    1984. "The Impossibility of Inductive Probability." *Nature* 310:433.
    1987. "The Demons of Decision." *Monist* 70:193–211.
Lewis, D.
    1980. "A Subjectivist's Guide to Objective Chance." In Jeffrey 1980.
    1986. "Postscripts to 'A Subjectivist's Guide to Objective Chance.'" In *Philosophical Pa-
    pers,* vol. 2 (New York: Oxford University Press).
Lindley, D. V.
    1982. "Scoring Rules and the Inevitability of Probability." *International Statistical Review*
    50:1–26.
Mackie, J. L.
    1963. "The Paradox of Confirmation." *British Journal for the Philosophy of Science* 13:
    265–277.
Maher, P.
    1992. "Diachronic Rationality." *Philosophy of Science,* forthcoming.
Maxwell, G., and Anderson, R. M., eds.
    1975. *Induction, Probability, and Confirmation.* Minnesota Studies in the Philosophy of
    Science, vol. 6. Minneapolis: University of Minnesota Press.
Mayo, D.
    1991. "Novel Evidence and Severe Tests." *Philosophy of Science* 58:523–552.
Mellor, D. H.
    1971. *The Matter of Chance.* Cambridge: Cambridge University Press.
Merton, R. K.
    1973. "The Normative Structure of Science." In *The Sociology of Science* (Chicago: Univer-
    sity of Chicago Press).
Mie, G.
    1914. "Grundlagen einer Theorie der Materie." *Annalen der Physik* 40:25–63.
Miller, D. W.
    1966. "A Paradox of Information." *British Journal for the Philosophy of Science* 17:59–61.
Miller, R.
    1987. *Fact and Method.* Princeton: Princeton University Press.
Milne, E. A.
    1948. *Kinematic Relativity.* Oxford: Oxford University Press.

Minkowski, H.
1908. "Die Grundgleichungen für die elektromagnetischen Vorgange in der bewegten Körpern." *Nachrichten von der königlichen Gesellschaft der Wissenschaften zu Göttingen*, pp. 53–111.

Molina, E. C.
1930. "The Theory of Probability: Some Comments on Laplace's *Théorie Analytique*." *Bulletin of the American Mathematical Society* 36:369–392.
1931. "Bayes' Theorem: An Expository Presentation." *Annals of Mathematical Statistics* 2:23–37.

Murray, F. H.
1930. "Note on a Scholium of Bayes." *Bulletin of the American Mathematical Society* 36:129–132.

Neyman, J.
1937. "Outline of a Theory of Statistical Estimation Based on the Classical Theory of Probability." In *A Selection of Early Statistical Papers of J. Neyman* (Berkeley: University of California Press, 1967).

Ni, W. T.
1972. "Theoretical Frameworks for Testing Relativistic Gravity. IV: A Compendium of Metric Theories of Gravity and Their Post-Newtonian Limits." *Astrophysical Journal* 176:769–796.
1973. "A New Theory of Gravity." *Physical Review*, D7:2880–2883.

Niiniluoto, I.
1983. "Novel Facts and Bayesianism." *British Journal for the Philosophy of Science* 34:375–379.

Nordström, G.
1912. "Relativitätsprinzip und Gravitätion." *Physikalische Zeitschrift* 13:1126–1129.
1913. "Zur Theorie der Gravitätion vom Standpunkt der Relativitätsprinzip." *Annalen der Physik* 42:533–554.

Norton, J.
1989. "Eliminative Induction as a Method of Discovery: How Einstein Discovered General Relativity." Preprint.
1990. "The Determination of Theory by Evidence: The Case for Quantum Discontinuity, 1900–1915." *Synthese*, forthcoming.

Osherson, D. N., Stob, M., and Weinstein, S.
1986. *Systems That Learn*. Cambridge: MIT Press.
1988. "Mechanical Learners Pay a Price for Bayesianism." *Journal of Symbolic Logic* 53:1245–1251.

Osherson, D. N., and Weinstein, S.
1989a. "Identifiable Collections of Countable Structures." *Philosophy of Science* 56:95–105.
1989b. "Paradigms of Truth Detection." *Journal of Philosophical Logic* 18:1–42.

Pais, A.
1982. *Subtle Is the Lord: The Life and Science of Albert Einstein*. Oxford: Oxford University Press.

Pearson, K.
1920a. "The Fundamental Problem of Practical Statistics." *Biometrika* 13:1–16.
1920b. "Note on 'The Fundamental Problem of Practical Statistics.'" *Biometrika* 13:300–301.
1978. *The History of Statistics in the Seventeenth and Eighteenth Centuries against the Changing Background of Intellectual, Scientific, and Religious Thought*. London: Charles Griffin.

Penrose, R.
1989. *The Emperor's New Mind*. Oxford: Oxford University Press.

Pitman, E. J. G.
1965. "Some Remarks on Statistical Inference." In J. Neyman and L. M. LeCam, eds., *Bernoulli, 1713; Bayes, 1763; Laplace, 1813* (New York: Springer-Verlag).

Popper, K.
1961. *The Logic of Scientific Discovery*. New York: Science Editions.

Popper, K., and Miller, D. W.
1983. "A Proof of the Impossibility of Inductive Probability." *Nature* 302:687–688.
1987. "Why Probabilistic Support Is Not Inductive." *Philosophical Transactions of the Royal Society* (London) A321:569–591.

Price, R.
1767. *Four Dissertations*. London: A. Millar and T. Cadell. 2nd ed., 1768.

Putnam, H.
1963a. "'Degree of Confirmation' and Inductive Logic." In Schilpp 1963. Reprinted in Putnam 1975.
1963b. "Probability and Confirmation." In *The Voice of America Forum: Philosophy of Science*, vol. 10 (U.S. Information Agency). Reprinted in Putnam 1975.
1975. *Mathematics, Matter, and Method*. Cambridge: Cambridge University Press.

Ramsey, F. P.
1931. "Truth and Probability." In R. B. Braithwaite, ed., *Foundations of Mathematics and Other Logical Essays* (London: Routledge and Kegan Paul). Reprinted in Kyburg and Smokler 1964.

Redhead, M. L. G.
1980. "A Bayesian Reconstruction of the Methodology of Scientific Research Programs." *Studies in the History and Philosophy of Science* 11:341–347.
1985. "On the Impossibility of Inductive Logic." *British Journal for the Philosophy of Science* 36:185–191.

Reichenbach, H.
1961. *Experience and Prediction*. Chicago: University of Chicago Press.

Rivest, R., Haussler, D., and Warmuth, M., eds.
1989. *Proceedings of the Second Annual Workshop on Computational Learning*. San Mateo, Calif.: Morgan Kaufman.

Robertson, H. P.
1962. "Relativity and Cosmology." In A. J. Deutch and W. B. Klemperer, eds., *Space Age Astronomy* (New York: Academic Press).

Rogers, H.
1987. *Theory of Recursive Functions and Effective Computability*. Cambridge: MIT Press.

Roll, P. G., Krotov, R., and Dicke, R. H.
1964. "The Equivalence of Inertial and Passive Gravitational Mass." *Annals of Physics* 26:442–517.

Rosenkrantz, R. D.
1981. *Foundations and Applications of Inductive Probability*. Atascadero, Calif.: Ridgeview.
1983. "Why Glymour Is a Bayesian." In Earman 1983.

Roseveare, N. T.
1982. *Mercury's Perihelion from Le Verrier to Einstein*. Oxford: Oxford University Press.

Russell, B.
1948. *Human Knowledge: Its Scope and Limits*. New York: Simon and Schuster.

Salmon, W.
1968. "Who Needs Rules of Acceptance?" In Lakatos 1968.
1973. "Confirmation." *Scientific American*, May, 75–83.
1975. "Confirmation and Relevance." In Maxwell and Anderson 1975.
1979. "The Philosophy of Hans Reichenbach." In W. Salmon, ed., *Hans Reichenbach: Logical Empiricist* (Dordrecht: D. Reidel).
1990. "Rationality and Objectivity in Science, *or* Tom Kuhn Meets Tom Bayes." In C. W. Savage, ed., *Scientific Theories*, Minnesota Studies in the Philosophy of Science, vol. 14 (Minneapolis: University of Minnesota Press).

Savage, L. J.
  1954. *Foundations of Statistics*. New York: John Wiley.
Schervish, M. J., and Seidenfeld, T.
  1990. "An Approach to Consensus and Certainty with Increasing Evidence." *Journal of Statistical Planning and Inference* 25:401–414.
Schervish, M. J., Seidenfeld, T., and Kadane, J. B.
  1984. "The Extent of the Non-conglomerability in Finitely Additive Probabilities." *Zeitschrift für Wahrscheinlichkeitstheorie* 66:205–226.
  1990. "State Dependent Utilities." *Journal of the American Statistical Association* 85:840–847.
Schick, F.
  1986. "Dutch Bookies and Money Pumps." *Journal of Philosophy* 83:112–119.
Schiff, L. I.
  1967. "Comparison of Theory and Observation in General Relativity." In J. Ehlers, ed., *Relativity Theory and Astrophysics. I: Relativity and Cosmology* (Providence, R.I.: American Mathematical Society).
Schilpp, P. A., ed.
  1963. *The Philosophy of Rudolf Carnap*. La Salle: Open Court.
Seelig, C.
  1956. *Albert Einstein: A Documentary Biography*. London: Staples Press.
Seidenfeld, T.
  1979. "Why I Am Not an Objective Bayesian: Some Reflections Prompted by Rosenkrantz." *Theory and Decision* 11:413–440.
  1986. "Entropy and Uncertainty." *Philosophy of Science* 53:467–491.
  1988. "Decision Theory without 'Independence' or without 'Ordering.'" *Economics and Philosophy* 4:267–290.
Seidenfeld, T., and Schervish, M. J.
  1983. "Conflict between Finite Additivity and Avoiding Dutch-Book." *Philosophy of Science* 50:398–412.
Seidenfeld, T., Schervish, M. J., and Kadane, J. B.
  1990. "When Fair Betting Odds Are Not Degrees of Belief." In *PSA 1990* (Philosophy of Science Assoc., East Lansing, Mich.) vol. 1.
Shafer, G.
  1982. "Bayes' Two Arguments for the Rule of Conditioning." *Annals of Statistics* 10:1075–1089.
Shapere, D.
  1966. "Meaning and Scientific Change." In R. G. Colodny, ed., *Mind and Cosmos: Essays in Contemporary Science and Philosophy* (Pittsburgh: University of Pittsburgh Press).
Shimony, A.
  1955. "Coherence and the Axioms of Confirmation." *Journal of Symbolic Logic* 20:1–28.
  1970. "Scientific Inference." In R. G. Colodny, ed., *The Nature and Function of Scientific Theories* (Pittsburgh: University of Pittsburgh Press).
  1988. "An Adamite Derivation of the Principles of the Calculus of Probability." In J. H. Fetzer, ed., *Probability and Causality* (Dordrecht: D. Reidel).
Silberstein, L.
  1918. "General Relativity Theory without the Equivalence Hypothesis." *Philosophical Magazine* 36:94–128.
Skyrms, B.
  1983. "Three Ways to Give a Probability Assignment a Memory." In Earman 1983.
  1987. "Dynamic Coherence and Probability Kinematics." *Philosophy of Science* 54:1–20.
Smith, A. F. M.
  1986. "Why Isn't Everyone a Bayesian? Comment." *American Statistician* 40 (no. 1):10.
Smith, C. A. B.
  1961. "Consistency in Statistical Inference and Decision." *Journal of the Royal Statistical Society*, B23:1–25.

Spielman, S.
   1977. "Physical Probability and Bayesian Statistics." *Synthese* 36:235–269.
Stigler, S. M.
   1982. "Thomas Bayes' Bayesian Inference." *Journal of the Royal Statistical Society*, Series A, 145:250–258.
   1986. *The History of Statistics*. Cambridge: Harvard University Press.
Suppe, F.
   1989. *The Semantic Conception of Theories and Scientific Realism*. Urbana: University of Illinois Press.
Suppes, P.
   1966. "A Bayesian Approach to the Paradoxes of the Ravens." In Hintikka and Suppes 1966.
Swinburne, R.
   1979. *The Existence of God*. Oxford: Oxford University Press.
Teller, P.
   1969. "Goodman's Theory of Projection." *British Journal for the Philosophy of Science* 20:219–238.
   1973. "Conditionalization and Observation." *Synthese* 26:218–258.
   1975. "Shimony's A Priori Arguments for Tempered Personalism." In Maxwell and Anderson 1975.
   1976. "Conditionalization, Observation, and Change of Preference." In W. Harper and C. A. Hooker, eds., *Foundations of Probability Theory, Statistical Inference, and Statistical Theories of Science* (Dordrecht: D. Reidel).
Teller, P., and Fine, A.
   1975. "A Characterization of Conditional Probability." *Mathematical Magazine* 48:267–270.
Thorne, K. S., Lee, D. L., and Lightman, A. P.
   1973. "Foundation for a Theory of Gravitation Theories." *Physical Review*, D7:3563–3578.
Thorne, K. S., and Will, C. M.
   1971. "Theoretical Frameworks for Testing Relativistic Gravity. I: Foundations." *Astrophysical Journal* 163:595–610.
Todhunter, I.
   1865. *A History of the Mathematical Theory of Probability from the Time of Pascal to That of Laplace*. London: Macmillan.
Van Fraassen, B. C.
   1980. *The Scientific Image*. Oxford: Oxford University Press.
   1983a. "Calibration: A Frequency Justification for Personal Probability." In R. S. Cohen and L. Laudan, eds., *Physics, Philosophy, and Psychoanalysis: Essays in Honor of Adolf Grünbaum* (Dordrecht: D. Reidel).
   1983b. "Theory Comparison and Relevant Evidence." In Earman 1983.
   1988. "The Problem of Old Evidence." In D. F. Austin, ed., *Philosophical Analysis* (Dordrecht: Kluwer Academic).
   1989. *Laws and Symmetry*. New York: Oxford University Press.
   1990. "Rationality Does Not Require Conditionalization." In E. Ullman-Margalit, ed., *The Israel Colloquium: Studies in History, Philosophy, and Sociology of Science* (Dordrecht: Kluwer Academic), forthcoming.
Von Plato, J.
   1982. "The Significance of the Ergodic Decomposition of Stationary Measures for the Interpretation of Probability." *Synthese* 53:419–432.
Watkins, J.
   1987. "Lawson on the Raven Paradox and Background Knowledge." *British Journal for the Philosophy of Science* 38:567–571.

Will, C. M.

1971a. "Theoretical Frameworks for Testing Relativistic Gravity. II: Parameterized Post-Newtonian Hydrodynamics and the Nordtvedt Effect." *Astrophysical Journal* 163:611–628.

1971b. "Theoretical Frameworks for Testing Relativistic Gravity. III: Conservation Laws, Lorentz Invariance, and Values of the PPN Parameters." *Astrophysical Journal* 169: 125–140.

1972. "Einstein on the Firing Line." *Physics Today* 25 (no. 10):23–29.

1974a. "Gravitation Theory." *Scientific American*, Nov., 25–33.

1974b. "The Theoretical Tools of Experimental Gravitation." In B. Bertotti, ed., *Experimental Gravitation* (New York: Academic Press).

1981. *Theory and Experiment in Gravitational Physics*. Cambridge: Cambridge University Press.

1984. "The Confrontation between General Relativity and Experiment: An Update." *Physics Reports* 113 (no. 6):345–422.

Williams, P. M.

1976. "Indeterminate Probabilities." In M. Prezlecki, K. Szaniawski, and R. Wojcicki, eds., *Formal Methods in the Methodology of the Empirical Sciences* (Dordrecht: D. Reidel).

Wittgenstein, L.

1961. *Tractatus Logico-philosophicus*. London: Routledge and Kegan Paul.

Worrall, J.

1978. "The Ways in Which the Methodology of Scientific Research Programs Improves on Popper's Methodology." In G. Radnitzky and G. Andersson, eds., *Progress and Rationality in Science* (Dordrecht: D. Reidel).

1985. "Scientific Discovery and Theory-Confirmation." In J. Pitt, ed., *Change and Progress in Modern Science* (Dordrecht: D. Reidel).

1989. "Fresnel, Poisson, and the White Spot: The Role of Successful Predictions in the Acceptance of Scientific Theories." In D. Gooding, T. Pinch, and S. Schaffer, eds., *The Uses of Experiment* (Cambridge: Cambridge University Press).

1991. "Falsification, Rationality, and the Duhem Problem: Grünbaum versus Bayes." To appear in a *Festschrift* for Adolf Grünbaum, to be published by the University of Pittsburgh Press and the University of Konstanz Press.

Zabell, S. L.

1988. "Symmetry and Its Discontents." In W. Harper and B. Skyrms, eds., *Causation, Chance, and Credence*, vol. 1 (Dordrecht: Kluwer Academic).

1989. "The Rule of Succession." *Erkenntnis* 31:283–321.

Index